VIVEK WADHWA AND ALEX SALKEVER

YOUR

HAPPINESS

WAS

HACKED

WHY TECH IS WINNING THE BATTLE TO CONTROL YOUR BRAIN, AND HOW TO FIGHT BACK

BK®

Berrett–Koehler Publishers, Inc.
a BK Business book

Berrett-Koehler Publishers, Inc.
1333 Broadway, Suite 1000
Oakland, CA 94612-1921
Tel: (510) 817-2277
Fax: (510) 817-2278
www.bkconnection.com

ORDERING INFORMATION

Quantity sales. Special discounts are available on quantity purchases by corporations, associations, and others. For details, contact the "Special Sales Department" at the Berrett-Koehler address above.

Individual sales. Berrett-Koehler publications are available through most bookstores. They can also be ordered directly from Berrett-Koehler: Tel: (800) 929-2929; Fax: (802) 864-7626; www.bkconnection.com.

Orders for college textbook/course adoption use. Please contact Berrett-Koehler: Tel: (800) 929-2929; Fax: (802) 864-7626.

Distributed to the U.S. trade and internationally by Penguin Random House Publisher Services.

Berrett-Koehler and the BK logo are registered trademarks of Berrett-Koehler Publishers, Inc.

Printed in the United States of America

Berrett-Koehler books are printed on long-lasting acid-free paper. When it is available, we choose paper that has been manufactured by environmentally responsible processes. These may include using trees grown in sustainable forests, incorporating recycled paper, minimizing chlorine in bleaching, or recycling the energy produced at the paper mill.

Library of Congress Cataloging-in-Publication Data
Names: Wadhwa, Vivek, author. | Salkever, Alex, author.
Title: Your happiness was hacked : why tech is winning the battle to control your brain, and how to fight back / by Vivek Wadhwa and Alex Salkever.
Description: First edition. | Oakland, CA : Berrett-Koehler Publishers, Inc., [2018] | Includes bibliographical references.
Identifiers: LCCN 2018003600 | ISBN 9781523095841 (hardcover)
Subjects: LCSH: Information technology—Social aspects. | Information technology—Psychological aspects. | Internet addiction.
Classification: LCC HM851.W3335 2018 | DDC 303.48/33—dc23 LC at https://lccn.loc.gov/2018003600

FIRST EDITION

26 25 24 23 22 21 20 19 18 | 10 9 8 7 6 5 4 3 2 1

Book produced by BookMatters; copyedited by Mike Mollett; proofread by Janet Reed Blake; indexed by Leonard Rosenbaum. Cover designed by Rob Johnson, Toprotype, Inc.

CONTENTS

*This book is dedicated to our editor Neal Maillet
and to his late son Aaron Cooper Maillet, who passed
away in a tragic accident just as this book was going to
press. There is no easy way to come to terms with such a
profound loss. May time bring Neal, Jacqueline, Brenna,
and Hilary a measure of healing and comfort.*

FOREWORD

I have been a professional technology investor since 1982, which has given me a front-row seat at the creation of the most exciting industries of the past thirty-five years, including personal computers, cellular phones, the Internet, and social networking. I was a mentor to Mark Zuckerberg during the early years of Facebook and at one time was a vocal advocate for the platform. I still love the Facebook service, but I believe that the company's advertising business model has created social, economic, and political damage that demands a national conversation, and possibly intervention. And Facebook is not alone: the problem is endemic to Google, Snapchat, Twitter, Slack, and most of the other major Internet platforms.

Internet platforms have revolutionized our lives, but only now are we beginning to see their dark side. Millions of adults lose productivity, sleep, and motivation through constant interruptions by technology that was supposed to make them more productive. There has been widespread coverage of the way Russian hackers exploited Facebook's architecture to interfere in the U.S. presidential election of 2016. Less well known is the way that use of Facebook influenced the vote on Brexit in the U.K., as well as other recent elections in Europe. There is mounting evidence that Facebook is also being exploited by allies of the government in Myanmar to make genocide of the Rohingya minority acceptable to that country's population. In the U.S., Facebook's advertising tools enable illegal discrimination in housing and violate the civil rights of innocent people. The Internet platforms themselves are particularly dangerous for children, who do not have tools to protect themselves. Snapchat Streaks and similar products on other platforms substitute addictive activities for the human interaction that is so fundamental to the emotional well-being of children. On top of that, lack of vigilance by the platforms has resulted in millions of children being exposed to inappropriate content.

The good aspects of Internet platforms are now being offset by flaws that are invisible to most users. All of this is possible because Facebook, Google, and other Internet platforms consciously addict their users in order to make their products and advertising more valuable. They combine propaganda techniques initially developed by the U.K. government in World War I with addiction strategies perfected by the gambling industry. They deliver two billion individually personalized channels on smartphones, the first media-delivery platform that is available to users every waking moment. The Internet platforms give users "what they want," creating filter bubbles that reinforce pre-existing beliefs in ways that make those beliefs more extreme and inflexible, causing many users to reject new information and even evidence.

It is ironic that tech platforms have joined illegal drug dealers in calling their consumers "users." As are many illegal drug users, technology platform users are addicted. Too many have lost control over their lives. Too many cannot help themselves, because they either don't know they are addicted or don't have the tools with which to break the addiction. At present, there is no organized effort to help them.

A handful of Silicon Valley leaders—mostly people like me who had once been involved with Facebook or Google—recognized this problem in 2016 and 2017, and started to speak out. Meanwhile, the founders and CEOs of many major technology companies limit use of these products by their children, even as they promote unrestricted use by everyone else. Similarly, the platforms talk about privacy but take every step imaginable to invade the privacy of their users. They talk about connecting people, but their products actually increase polarization, isolation, and loneliness.

We are at a crossroads. In 2016, the tech industry could reasonably claim to be unaware of the problems pervading advertising-supported Internet platforms. That is no longer the case. Policy makers in Washington and around the world increasingly recognize that the promise of always-on technology has given way to a dystopian present. The time has come for "users" to get involved and to push back on platforms that are causing them harm in the pursuit of profits.

Your Happiness Was Hacked is a really important and

timely book. Not only is it the first on this topic by people who have spent their careers in the tech industry, but it also combines analysis of the problem with thoughtful prescriptions. It will not be easy to fix what is wrong with the major Internet platforms and our relationship to them, but the first step is to present the facts and foster a conversation about where to go from here. Vivek and Alex have taken that critical first step. They have surveyed the pioneering work being done by my partner Tristan Harris, by James Williams, and by many others, and distilled it into the book you are reading. There will be many more books about this issue, but this a great place to start.

Roger McNamee

PREFACE

Technology Overload Is Personal

Technology has given us so many gifts. Any information we need, Google lets us find within seconds. Facebook, Instagram, and Snapchat let us share our lives with distant friends and family. Our smartphones can be our running coaches, our libraries, and our meditation gurus. We no longer need to wrestle with paper maps; smartphones read detailed directions to us aloud while mapping the routes on their screens, even quickly rerouting us should we diverge from the plotted course. Uber and Lyft have made summoning a car as simple as pressing a button. Amazon can deliver ordered items within a day (and, in some cities, within two hours). Netflix streams movies to our screens for less than the cost of going to a single film at the cinema.

In the workplace, technology has forever altered our lives. E-mail allows us to communicate instantaneously and to have a permanent searchable record of our work. Slack, Facebook Messenger, and other instant-messaging applications let us chat and share files with work colleagues, and they build virtual watercoolers around which remote workers can gather to share stories, jokes, or GIFs.

When we create presentations or need information, we can sift through millions of available (and often free) online images. Or we can watch videos that teach us new skills for nearly any task—from relighting a water heater's pilot flame to using the most popular computer programs for artificial intelligence (AI). We get nearly all of the news we want, at any time, for free.

Traveling on planes, we face flat-panel displays that let us flick from channel to channel or from movie to movie, keeping boredom at bay. We ride on elevators facing televisions broadcasting the news and weather, just in case we were unhappy about wasting the 30 seconds ascending or descending. Dynamic digital billboards now turn roadsides, bus stops, and city streets into carousels of capitalism. And virtual reality promises endless fully immersive adventures, enabling any of us to travel the world without moving from our chairs. The wonders never cease.

Yet a growing volume of research finds that Americans are unhappier now than they have been at any time in the past decade—and are becoming unhappier.[1]

Psychologists raise the alarm over an epidemic of loneliness consuming society.[2] Rates of teenage suicide are rising, and today's teenagers are less happy than teenagers of previous generations.[3] They are also less likely to leave the house, hold a job, and do things that were once rites of passage.[4] Smartphone addiction has made distracted driving epidemic; nearly 3,500 people died and 391,000 were injured in vehicle accidents involving distracted drivers in 2015, and such accidents are becoming more common.[5]

Our stores of empathy are shrinking, and narcissism is becoming normal, both trends being potentially attributable to pervasive technology.[6,7]

Obsessive use of social media enables constant unhealthy comparisons with the seemingly perfect lives of those we see in our social-media feeds—even when we consciously know that their lives are less than perfect. More than one-third of the U.S. population gets less than the recommended minimum seven hours of sleep a night, with many millions getting less than six hours, and some of the best sleep researchers in the world consider incessant exposure to technology a likely leading cause. Most smartphone owners, fearing being away from their devices, sleep with their devices within arm's reach.[8] Naturally, they also respond to e-mails and social-media alerts when they wake up with their phones at their sides, a behavior no one thinks is healthy. Meanwhile, a growing body of research suggests that late-night exposure to the intense blue light emitted by most computer and smartphone screens impairs production of melatonin, a chemical essential to sound sleep.

From texts to tweets to e-mail newsletters to binge-watching TV series such as *Orange Is the New Black*, so many things demand our attention. We are inundated with red circles and alerts and sounds, all designed to tap deep into our brains and hijack the neural pathways that enabled our ancestors to detect threats and thereby survive. What should serve us as primal alarm systems have left us trapped instead in a downward spiral of anxiety and discontent.

We know that uncontrolled consumption of technology

is increasingly diverting us from our intentions, but we seem unable to stop. Research subjects even choose to receive electric shocks rather than be left alone with their thoughts and without any technologies.[9] The very engineers who built the devices that hold us rapt now express misgivings about what they have wrought (sending their own children to technology-free schools and restricting screen time at home), and the creator of the Facebook Like button now has his personal assistant use parental controls to prevent him from downloading apps to his phone.[10]

Even worse, some of the smartest people in the world are using powerful artificial-intelligence technologies specifically to devise ever newer and more effective ways to hold our attention.[11] We are collectively in the throes of a massive, harmful addiction that is the signature social issue of our time. This technology addiction is increasingly removing us from the direct experience of life, and that is consequently robbing us of our sense of peacefulness, security, stillness, and ease with ourselves. More cogently, our tech addiction has made it much harder for us to sit still or even to simply pay attention. The mechanism of this addiction is the steady, iterative diminution of our choices. This reduction of choice is a gentle slope. Like the frog boiling slowly in water, we spend increasing periods each day on our devices or interacting with technology, and our range of actual choices narrows.[12] This is not to say that we're consciously aware of such limits. To the contrary, we imagine we have never before had such a bounty of ways to amuse ourselves, learn, research, and consume information.

And it's true that we also benefit from this newfound digital store of knowledge. We can find forecasts of tomorrow's weather anywhere on the globe. We can quickly book flights or reserve tables at restaurants. We can snap pictures of our wage forms for software to convert into simple tax returns. On our phones, we can track the locations of our loved ones, and communicate in real time when we are late for appointments. And if we're involved in car accidents, we have phones with which to call for help—or applications that automatically detect that we have been in an accident.

But increasingly the choices we make are subtly (and not so subtly) manipulated by the makers of our technology in ways intended to promote the makers' profit over our individual and collective well-being.

In this book, we aim to help you understand why and how technology is making us so unhappy. And we correlate the rising use of smartphones, e-mail, social media, and other modern technologies with increasing angst, suffering, loneliness, and unhappiness. We analyze the scientific literature on how technology affects our lives. And we suggest what you can do about it.

Both of us, Vivek and Alex, came to write this book because we feel strongly about the negative effects that technology can have on our lives. Each of us has felt these effects acutely in recent years.

Neither of us hates technology. We both love it. And we could not imagine what our lives would be like without the massive benefits technology has provided to the world. We have made our careers in the technology field.

But as parents and spouses, as managers and entre-
preneurs, and as people, we have felt a growing unease
with technology over the past decade as it has become
more deeply embedded into our day-to-day existence. As
we shared the idea for our book with others, every single
person we spoke to felt what we were feeling: it's a problem
that affects our lives hugely.

A growing body of scientific evidence finds significant
negative side effects of many of the ways we use technology
and our habits in using the Internet, our smartphones, and
nearly all other digital formats. This book will help you recognize the scope of the problem: how technology's many
tentacles constrain and consume us in ways we fail to recognize. It describes how a form of techno-quicksand sucks
us in and reduces our satisfaction at work and at home,
puts us at mortal risk on the roads, and invades our most
intimate moments to weave an unhealthy web of compulsion and dependency. It employs anecdotes and scientific
research, and analyzes the ways in which companies, experts, scientists, and well-informed individuals are creating healthier relationships with technology and attempting
to recover their equilibrium and their choices.

Ultimately, we hope to show how you can use a series
of strategies and skills to build a better, more fulfilling life,
one that includes both technology and happiness.

Turning the clock back is neither a realistic nor a desired
option for most of us. We *like* Netflix. We *rely* on e-mail. We
don't really *want* to read a paper map. FaceTime is a great
way to stay close to people we care about. Expensify has

taken a lot of the pain out of filing expenses. And online shopping is incredibly convenient. What none of us bargain for are the convenience's hidden costs, increasingly compromising our day-to-day experience and our relationships.

Our society needs to ensure that the benefits of technology use outweigh the downsides and that we allow technology into our lives only on our terms. Otherwise, we risk a dystopian future in which we are slaves to our devices; in which we allow the very things that make being human so meaningful to drown in the noise of a million dopamine signals arising from alerts, social-media posts, beeps, rings, and notifications. Without being mindful in our technology use, we face a future of endless distraction and inattention that no one wants to endure.

Some of the urgency of the warnings about technology comes from acknowledgment of a stark reality: that the current generations may be the last who remember a life before this technology invasion overwhelmed us. Children born today will see the way we interact with our technology— staring at smartphones in the presence of crying children, interrupting deep thinking and writing projects for chatter on Slack, replying to texts as we drive—as the norm and as the only way that things *can* be. It is our responsibility to reshape this narrative and, as grandiose as it sounds, make technology safer not only for our children but for all generations to come.

Introduction

Alex Almost Kills a Pack of Cyclists

On a cloudy morning several years ago, Alex was driving on Highway 1 in Marin County, California, a serpentine road along the ocean cliffs. His mind was elsewhere. His company was about to make an urgent product and deal announcement in the week ahead, and the fallout had erupted into a weekend of back-and-forth rapid-fire messages. The entire senior executive team was included on e-mail threads and texts, and Alex felt that he was expected to reply quickly.

The iPhone mounted on the dashboard of the car kept buzzing. Alex knew how dangerous it is to look at a phone while driving, let alone while driving on this stretch of highway, but he couldn't stop his hands from picking up the phone to snatch pieces of messages whenever the curves briefly abated and the road straightened out. He knew well that he could have chosen instead to stop at a pullout. Even so, he kept on driving.

On a straight section of highway, as he was furtively tapping a reply, a sixth sense told Alex to look up. What he saw, less than fifteen feet away, was a pack of cyclists in bright red clothing, frantically pedaling up the steep grade. Alex hit the brakes, and the car skidded to a rapid stop. As Alex sat in his car, heart thumping, he realized that the cars behind him were honking. It was a narrow road, and on this stretch of it, only a narrow guardrail separated the cyclists from the cliff. Two seconds longer, and he would have hit the group, injuring them and potentially pushing them off the cliff to their deaths. Had that happened, Alex realized that he could well have ended the lives of the cyclists and scarred those of their families. Children might never have seen their mothers and fathers again. More selfishly, Alex also might have lost his job, put his own family under incredible stress, and forever changed his life. It was an utterly stupid, inexcusable act. It was also an utterly normal and common one: the vast majority of drivers who bring smartphones into the car interact with them while driving.[1]

Had Alex waited for twenty minutes to reply instead from his final destination, would it have mattered? He knew that it wouldn't have. Yet the pull was so strong and the risk so abstract that Alex, normally a clearheaded and responsible person, made a bad decision—and avoided unthinkable consequences by mere seconds.

Over the years, Alex had felt a growing unease over how his relationship with technology was influencing his behaviors. As a child, a teen, and later a university student, he could read a book or write on topics for hours on end. Then along came the web and e-mail, invaluable tools for a writer trying to build a career as a freelance writer and later as an editor at *Businessweek*. So Alex came to rely heavily on both tools to help him more efficiently locate information and talk with sources. But the habit of checking e-mail gradually became an unhealthy compulsion. Over time, from checking it every few hours, he came to check it hourly, and then every fifteen minutes. After all, he told himself, he never knew when an editor was going to e-mail with a request or when a source might respond to a question. He was in the news business.

One day, Alex realized that he *needed* to be connected to the Internet in order to write at all; he just felt strange when not connected. When connected, he could, as he saw it, write and research at the same time. But this also enabled him to continuously check e-mail and social media and to surf the web, diving down rabbit holes of useless information that popped up in his searches. Always a fan of notifications, Alex loved to be constantly accessible to colleagues and clients.

That said, when he was on vacation, Alex found it hard to slow down and unwind. He felt antsy when not connected, and connecting to airport Wi-Fi after a long journey became a quasi-religious experience. Once, he had laughed at the passengers who checked e-mails as soon as the wheels hit the runway; now, he had become one of them. His parents, his wife, and his children had all become used to the fact that Alex never really took his vacations—at least, not from the Internet.

During one of those vacations, on a beautiful island off the coast of Massachusetts, Alex decided to tally up how he was spending his time on line and how much of that time was going toward work. The catalyst for this was an innocent question from his son: did Alex have time to go to the beach that day, or did he need to keep doing work on his computer? The question had struck home. Alex was choosing to sacrifice precious moments with his family that he would never recover—and the memories of one another that he and the children would never have—for the sake of time on the Internet e-mailing and doing research for work. He suspected that the "work" was probably less than 50 percent of the time he was spending; that checking e-mails and reading news articles took him on a wandering path of distractions that stole his time.

So Alex got a notepad and, every thirty minutes, wrote down what he had done in the previous half hour. At the day's end, he tabulated how he had spent his time on the computer. He found that less than one-third of his time on line actually went to work tasks; the rest was spent in vapid minutes and hours of surfing, replying to e-mails, and doing other things that didn't need to be done on a beautiful summer day while his children were at the beach. Technology, he concluded, had turned him into the kind of person he did not want to be. He vowed to gain control of the monster.

Vivek Nearly Dies from E-Mail Withdrawal

To say that Vivek nearly died from e-mail withdrawal overstates the case, of course, but only by a little. Vivek grew up programming computers and immigrated to the United States to work in technology. As he ascended through the ranks at large financial institutions, and as the Internet grew in importance, he launched two start-ups and took one public. A natural networker, Vivek used technology to build a massive web of friendships and connections across business, media, and government. Maintaining that web of connections, however, took a considerable amount of energy.

Vivek's hometown paper published a full-page paean—titled "Viva, Vivek!"—to Vivek's relationships with his employees. Behind the scenes of this success, though, even as he juggled the tasks essential to managing a growing start-up with two hundred employees and closing multi-million-dollar deals, Vivek was spending ever more time feeding his network. And juggling all of this meant—he thought—staying constantly connected.

On a vacation cruise with his family to Cancun, Mexico, Vivek felt compelled to check his e-mails and his texts. His company was going through a difficult patch because of a downturn in the economy, and Vivek consequently felt distressed and miserable, even on vacation. What Vivek wanted, first and foremost, was access to e-mail so he could know what was going on and not miss anything. His wife, Tavinder, tried to tell him to slow down, not to worry, and to relax. He knew that he shouldn't check his e-mails. And in fact he couldn't: compounding his stress and frustration, the ship's computer systems weren't working.

Then Vivek started to feel chest pains. At first he

ignored them. As he climbed the pyramid of Chichén Itzá, in the Mayan ruins on Mexico's Yucatán Peninsula, the pains became increasingly severe, and he began to feel nauseous. The views were stupendous. People dreamed for their whole lives of visiting this location and walking up these steps. Yet, amid the majesty of one of the greatest civilizations ever, Vivek's focus was his wish to connect to the Internet.

On the flight home, the chest pains and nausea turned into a shooting electric current in his left arm, and Tavinder insisted he go to the doctor. Even then, Vivek said he needed to first go home to check his e-mail.

Fortunately, Tavinder prevailed; once they landed, she drove him directly to the hospital at the University of North Carolina. Vivek blacked out as he entered the emergency room, and sat propped up in a wheelchair as they registered him. His next memory was of waking up after lifesaving surgery. Had he waited another hour or two, his doctors told him, Vivek would have been dead; none of his e-mails would have mattered. Over the course of the cruise and on the flight home, he had been having a massive heart attack, which caused permanent injury to his heart.

It is impossible to precisely apportion the blame that e-mail and other technologies share for this, but Vivek is sure that the stress of feeling the need always to be digitally connected played a major role in his heart attack. The ceaseless need to feed the technology monster had subverted Vivek's awareness of the need to properly care for himself.

Vivek recovered and got off the corporate and start-up treadmills. He changed professions, from technology CEO to academic professor and researcher. His life goal became to educate and inspire others to make the world a better place. He gave up the pursuit of initial public offerings (IPOs) in favor of the pursuit of knowledge. He also began to learn about mindfulness. He started meditating, exercising, and hiking. It may sound clichéd, but he had realized that the old way of life—one of technology-induced stress—would kill him.

Though he now viewed technology with some caution, Vivek remained enthralled with its remarkable potential. Technology wasn't entirely bad, he knew. He believed that it had the potential to solve the world's greatest problems: hunger, thirst, lack of shelter, disease. In his native India, for example, technology was improving the lives of hundreds of millions of people by letting them communicate, giving them access to financial services, and making health care more affordable. Despite his love–hate relationship with technologies that demand attention (social media and e-mail), he knew that he had neither the desire nor the ability to entirely stop using them. He needed Twitter, Facebook, and LinkedIn to communicate with a broad group of followers around the globe, who even now share ideas with him and connect him with interesting people along the way. He wasn't about to give up e-mail and return to snail mail.

Still, Vivek recognized a building tension, a conflict with the happiness and mindfulness he felt when he took a break from technology on his hikes in nature or on vacations without smartphones. That conflict, he realized, reflected a false choice.

Vivek began to take note of the various ways in which technology was separating him from the people he cared about. He noted that he often sent text messages to his sons instead of speaking to them, even if they sat in the next room. He noted that he spent less time with old friends and felt satisfied sending them e-mails. Broadly, he found that he had begun to avoid speaking on the phone unless it was entirely necessary. In fact, he sensed that technology had made him less patient and less willing to wait: less empathetic.

How Technology Hacks Our Happiness

In this book, we dive deeply into what caused the unhealthy behaviors that became our normal state of existence for many years. We are both seasoned technology executives who have been immersed in technology since our earliest years. We both spent time programming computers in our youths in the early days of PCs. Both of us were early adopters of the Internet. Vivek built two software start-ups and worked as both a programmer and a senior technology executive at a major investment bank. Alex began his career in journalism covering technology before going to work for a series of technology start-ups and companies, including one, Mozilla, that develops browsers and seeks to maximize consumer consumption (as do all browsers and nearly all phone and web applications).

We have both been wary of the impacts of technology on our lives yet helpless to control our relationship with it—which included compulsive checking of social media or e-mail, texting while driving, and watching specific queries on Google or YouTube digress into random excursions across the Internet. And in the back of our minds, we have both started to wonder whether what others perceive as our diminished patience and what we perceive as diminished empathy may reside in subtle but critical changes in the way our brains function as a result of our constant immersion in technology. (And research findings that use of the technology leads to changes in physical brain structure—see below—give such concerns a strong basis.)

We know that we may come across as grumpy quasi-Luddites lecturing millennials and Generation Z on how messed up their lives are and how technology is destroying their generations. That isn't our intent. In many realms, as we acknowledge, technology has made our lives significantly better and emotionally richer by giving us amazing, unprecedented ways to connect. Alex and Vivek both continue to use technology to communicate with their children and spouses. Truth be told, we have both been hypocrites, simultaneously lecturing our children and others on the negative impacts of technology (and in Alex's case, restricting its use in his home) while using technology in the same destructive fashion we speak against because its value remains undiminished in our eyes.

Rather, we want you, our reader, to think about this: technology is not always a benign, innocuous device with a screen that we can turn on and off when we want to. The companies that make technology—software and hardware—have their own reasons to command our attention, and their means of doing so are not confined to traditional tools of manipulation. The artificial intelligence they deploy seeks increasingly to surreptitiously guide our movements and thoughts, outsmarting us and influencing us in subtle ways to do the companies' bidding (Click on more ads! Like more posts! Don't leave, ever!). These companies employ brilliant mathematicians and data scientists to persuade us to use their offerings—generally meaning spending significant time and attention.

We are encouraged that so many in the tech sector seem to be waking up to the dangers long foreseen by visionaries such as Steve Jobs and Bill Gates (both of whom severely restricted their children's use of technology even as their companies sold their products aggressively to schools and children). But the current corporate demand for our increasingly scarce attention, in what has been dubbed an "attention economy," is designed to translate our time into income for corporate coffers. This is why Facebook, Google, and all other companies that traffic in messaging, social networking, browsing, and similar activities measure their success in amount of time spent per user, or in the number of actions a user takes (Likes, searches, clicks, tweets). When that number rises over time, investors are usually happy. When it falls, or even when it rises too slowly, someone's job is at risk—laying even more painfully bare the reality that technology companies are primarily (no surprise) in it for the money. Of course, we knew that. But we also listened to high-minded language about "connecting the world" and "organizing all the world's information." For a long time, we gave those companies a free pass. It's time for us to wake up and examine, gimlet eyed, every interaction we have on line and to think hard about how and when technology commands our attention—and, most importantly, to what end.

In seeking to reduce our choices, attention-economy companies limit our ability to choose for ourselves. This is how they control the game and tilt the scale in their favor, and this is why the news and information appearing to us on social-media sites exclude information that might challenge our worldviews. This is also why search results today favor larger companies and stores over the family-owned neighborhood stores that sponsor our local sports teams, pay property taxes, and give back directly to our communities. Though the big chains can pay for our attention, the small stores can't afford to. And so technology is nudging us toward choices with long-term implications for our communities—which, with every click, we remake in the image that the tech giants desire.

As well as affecting our immediate relationships with technology, these restrictions in choice have secondary and tertiary impacts. A growing body of research shows that technology exposure diminishes empathy.[2] A strong correlation has also been found between the increase in use of technology and a reduction in book reading.[3] Given the nature of our world and of this kind of technology, that trend may seem logical and benign. But reading books has long been associated with numerous positive human outcomes in education and in life. So anything that leads to a reduction in book reading should be weighed carefully for its positive and negative effects over the long term. A growing body of evidence suggests that people remember and learn more from offline reading than from reading on electronic devices.[4]

The technology we use may also be changing the physical structure of our brains.[5] Geospatial perception, for instance—map-reading ability and spatial awareness—may be taking a hit from the nearly universal adoption of GPS in smartphones and other devices. Early evidence indicates that this may result in a reduction in the size of the hippocampus in the brain, which plays critical roles in memory formation, learning, and happiness. We simply don't yet know what all the long-term impacts of this change will be.

The brain's plasticity makes it amazingly adaptable, and this adaptation may well help us deal with modern life. It may free up the hippocampus for tasks more urgent than those we can outsource to Google Maps. But choosing not to think about such matters is a way of ceding our choice and free will. Much as Daniel Kahneman considers two types of happiness and two types of thinking in his seminal book *Thinking, Fast and Slow*, we need to consider the impacts of technology in multiple ranges: the immediate direct impacts, the immediate or near-term secondary impacts, and the longer-term impacts.[6]

Fitness trackers are a case in point. A few years ago,

these devices were almost universally hailed as a simple, effective way for technology to drive healthy behaviors. Creating a reward structure for activity and movement, and taking advantage of the same psychological incentives (discussed in chapter 1) that drive us to continually track our social media feeds to constantly track our step counts should be good for humankind, right?

It turns out that humans adapt fairly quickly to fitness trackers and compensate for the step counts in unforeseen ways. A 2016 University of Pittsburgh study put 470 people on a low-calorie diet to lose weight.[7] Some of the participants were given fitness trackers, and others were not. After two years, participants who had worn fitness trackers had lost less body mass than those who had not. What had happened? People who used fitness trackers justified eating more on days when the fitness trackers recorded more exercise. Relying on external technological feedback in lieu of hunger signals, users of these devices ate more than they otherwise would have.

Another study, this time using fitness trackers and calorie counters in conjunction, gave mixed feedback on the devices.[8] Some participants reported improved restraint in eating, but others reported that the use of activity-tracking and calorie-counting technology increased symptoms of eating disorders.

This is a recurrent theme in this book: some people handle technology better than others, which is why a one-size-fits-all approach is wrong for technology usage. As these examples show, our relationship with technology is complicated, and the effects can rarely be seen in black and white. What is different and more urgent now is the rapid adoption that the newest technology systems have enjoyed. Fitness trackers went from fringe to nearly mainstream in less than five years in the developed world. Smartphones had become mainstream in slightly more than a decade, and the Internet took longer than that. Smart speakers, such as Google Home and Amazon's Alexa-powered Echo, are being adopted more quickly than smartphones or fitness trackers; these speakers bring an entirely new way for us to interact with technology.[9] Today, we may be on the cusp of embracing and entering another rapid-adoption cycle: that of virtual reality and augmented reality (VR/AR).

More than any other human–computer interface introduced to date, VR/AR, with which we will interact through multiple senses, has the potential to overwhelm our defenses and become highly addictive. The web between technology and our senses is tightening. Even now, Elon Musk, perhaps the greatest technology entrepreneur of our time, is building products to directly link our brains to technology, bypassing fingers, voice, and other physical command structures.[10] These are uncharted waters, and we urgently need an understanding of the three ranges of technology impact on humans (immediate direct effects, immediate to short-term secondary effects, and longer-term effects), and of the positives and negatives of every new rapid-technology cycle.

We, Vivek and Alex, believe that this boils down to a question of conscious choice. Technology that augments our choices, or that we use in such a way as to broaden them, will augment our free will and our fulfillment. Technology that surreptitiously reduces our choices, that seeks to constrain us rather than vice versa, will limit and reduce them. We also keep in mind the paradox of choice: the poverty of riches that is yet another facet of modern technology (for more on which, see chapter 3). Addressing the importance of choice by simply increasing the number of options to choose from doesn't acknowledge the way our psyche functions. Rather, we frame the goal as maximizing both *conscious* choice and our ability, as thinking citizens of the world, to *define* our choices, define our lives, and, in doing so, regain control, living intentionally and so becoming happier and more productive.

That is the goal of this book: to build a different way for us to think about technology.

In our previous book, *The Driver in the Driverless Car: How Our Technology Choices Will Create the Future*, we posed three questions to ask of any new technology:[11]

♦ Does it have the potential to benefit everyone equally?

♦ What are the risks and the rewards?

♦ Does it foster autonomy or dependency?

In this book, we take the next step and ask questions (and provide some answers) as to how we can regain control. Our society must learn to maintain our relationship with technology on terms that make it, on balance, a positive set of tools, maximizing the wonderful things technology can do for us and minimizing the harms it inflicts. All of us, millennials and baby boomers alike, can benefit from a healthier relationship with technology. Our aim in this book is to help bring that about.

How Technology Removes Our Choices

The Tricks and Tactics Tech Uses to Control
Our Actions and Stoke Addictions

If you use Google to search for "Italian restaurant," you are likely to see a small box at the top of the screen with a few results below a map. The positioning is significant: viewers are significantly more likely to click on those results than on anything else on the page, much as shoppers are more likely to pick up products from shelves at eye level in supermarkets than from higher and lower shelves.[1,2] But whereas in the physical world this limitation primarily affects our shopping experience, in the online and technology worlds, this algorithmic and sometimes intentional selection affects every subsequent thing that we see or do on that page—and far beyond it. The menu is the interface that controls the manner of engagement and sets limits on it, and the way menus are layered can radically alter the way we behave with technology.

For example, on iPhones Apple has an important—to Alex, critical—feature: the toggle that wipes in-app advertising identifiers that app makers can use to analyze and track users. Unfortunately, Apple places that feature deep in the menu: three layers deep. As a result, few people use it, even though regularly using the feature might significantly benefit their privacy by making it much harder for companies to track their behavior in smartphone apps. (The industry would say that using it would lead people to have less personalized and less useful experiences, which is certainly true; there is always a trade-off.)

Apple has in general taken a strong leadership position in protecting the privacy of its customers—by minimizing storage of customer data and by designing systems such as Apple Pay to present fewer opportunities for third parties to access and potentially intercept those data. But its placement of that single toggle deep in the weeds on the iPhone illustrates how decisions by product makers influence our freedom of choice and our relationship with technology. By clearing that identifier regularly, phone users would wipe away some of the capabilities of application developers to accurately target and personalize in-product offers, e-mails, and other entreaties to further guide or limit our choices and set the agenda for us.

Another example is the ability to set notifications in the iPhone. Apple does not allow us to make global changes to all the notification settings of our apps. This means we must go through, app by app, and set notification settings. Sure, we can turn them all off by putting our device in "Do Not Disturb" mode. But that is a clumsy fix. Apple's menu design for managing notifications reduces our choices and not necessarily to our advantage (which seems odd from Apple, a company that has become dominant precisely by simplifying technology).

As a number of thinkers in this field, led by former Google design ethicist Tristan Harris, explain, menus also frame our view of the world. A menu that shows our "most important" e-mails becomes a list of the people we have corresponded with most often recently rather than of those who are most important to us. A message that asks "Who wants to meet for brunch tomorrow?" goes out to the most recent group of people we have sent a group text to, or to preset groups of friends, effectively locking in these groups and locking out new people we have met. On the set of potential responses to e-mail that Google automatically suggests in its Inbox e-mail program, we have yet to see "Pick up the phone and call this person" as an option, even if, after a heated e-mail exchange, a call or a face-to-face conversation may well be the best way to communicate and to smooth the waters.

A feed of world news becomes a list built by a nameless, faceless algorithm of topics and events the system decides interest us. It limits our choice by confining it to options within a set of patterns deriving from our past consumption history, and this may or may not relate to our immediate needs or interests. Unfortunately, no one has yet developed an effective algorithm for serendipity.

From the start of the day, a feed of what we missed on Facebook or Twitter as we slept presents us with a menu of comparisons that stokes our fear of missing out (FOMO). This is so by design. However benign its intent, its effect is to significantly limit our frames of reference and our thinking.

A Slot Machine in Our Pocket

In May 2016, Tristan Harris published an influential essay titled "How technology is highjacking your mind—from a magician and Google design ethicist," describing the many ways by which smartphones suck people into their vortex and demand constant attention. Harris traced the lineage of (both inadvertent and intentional) manipulation common in the design of technology products directly to the numerous techniques that slot-machine designers use to entice gamblers to sit for hours losing money.[3]

Inspired by Harris and other advocates of more-mindful technology product design, a small but growing Silicon Valley movement in behavioral design is advocating greater consideration of the ethics and the human outcomes of technology consumption. (After leaving Google, Harris launched a website, Time Well Spent, that focuses on helping people build healthier interactions with technology.)

Harris, New York University marketing professor Adam Alter, and others have criticized the various techniques that product designers are using to encourage us to consume ever more technology even to our own clear detriment. Tightly controlling menus to direct our attention is one common technique (one that is not as easily available to offline businesses). For his part, Harris suggests that we ask four questions whenever we're presented with online menus: (1) What's not on the menu? (2) Why am I being given these options and not others? (3) Do I know the menu provider's goals? (4) Is this menu empowering for my original need, or are the choices actually a distraction? We assure you, once you start asking these questions, you will never look at the Internet or at software applications in the same light again!

Another technique, alluded to in the title of Harris's slot-machine article, is the use of intermittent variable rewards: unpredictability in the rewards of an interaction. The first behaviorist, psychologist B. F. Skinner, introduced this concept with his "Skinner box" research.[4] Skinner put rats into boxes and taught them to push levers to receive a food pellet. The rats learned the connection between behavior and reward quickly, in only a few tries. With further research, Skinner learned that the best way to keep the rats motivated to press the lever repeatedly was to reward them with a pellet only some of the time—to give intermittent variable rewards. Otherwise, the rats pushed the lever only when they were hungry.

The casinos took the concept of the Skinner box and raised it to a fine art, designing multiple forms of variable rewards into the modern computerized versions of slot machines. Those machines now take in 70 to 80 percent of casino profits (or, according to an industry official, even 85 percent).[5,6] Players not only receive payouts at seemingly random intervals but also receive partial payouts that feel like a win even if the player in fact loses money over all on a turn. With the newer video slots, players can place dozens of bets on the repetition of a screen icon in various directions and in varying sequence lengths.

Older mechanical slot machines displayed three reels

and one line. Newer video slot machines display digital icon grids of five by five or more. This allows for many more types of bets and multiple bets in the same turn. For example, the player can bet on how many times the same icon will appear in a single row, how many times it will appear on a diagonal, and how many times it will appear in a screen full of icons, all in one turn. This allows players to win one or more small bets during a turn and gain the thrill of victory, albeit that in aggregate they lost money on their collective bets for the turn. The brain's pleasure centers do not distinguish well between actual winning and the techniques that researchers call *losses disguised as wins* (LDW).[7] The machines are also programmed to highlight near misses (nearly enough of the right numbers), since near misses actually stimulate the same neurons as real wins do.[8]

Machine designers use myriad other clever sensory tricks—both visual and auditory—to stimulate our neurons in ways that encourage more playing. As explained in a 2014 article in *The Conversation*, "Losses disguised as wins, the science behind casino profits,"

> Special symbols might be placed on the reels that provide 10 free spins whenever three appear anywhere within the game screen. These symbols will often make a special sound, such as a loud thud when they land; and if two symbols land, many games will begin to play fast tempo music, display flashing lights around the remaining reels, and accelerate the rate of spin to enhance the saliency of the event. When you win these

sorts of outcomes you feel as though you have won a jackpot; after all, 10 free spins is 10x the chances to win big money right? The reality is that those 10 free spins do not change the already small probability of winning on any given spin and are still likely to result in a loss of money. For many games, features such as this have entirely replaced standard jackpots.[9]

What helps these techniques entice humans to keep playing is that our brains are hard wired to become easily addicted to variable rewards. This makes sense when you think that finding food in prehistoric, pre-agricultural times was a perfect example of intermittent variable rewards. According to research by Robert Breen, video-based gambling games (of which slots represent the majority) that rely on intermittent variable rewards result in gambling addiction three to four times faster than does betting on card games or sporting events.[10]

Smartphones were not explicitly designed to behave like slot machines, but their effect is nearly the same. As Harris writes,

> When we pull our phone out of our pocket, we're playing a slot machine to see what notifications we got. When we pull to refresh our email, we're playing a slot machine to see what new email we got. When we swipe down our finger to scroll the Instagram feed, we're playing a slot machine to see what photo comes next. When we swipe faces left/right on dating apps like Tinder, we're playing a slot machine to see if we got a match. When we tap the [red badge showing us the number

of notifications in an app], we're playing a slot machine to [see] what's underneath.[11]

Through this lens we can see how many actions deeply embedded in the technology we use are acting as variable rewards systems, and when we look at the technology in our lives, we can find intermittent variable rewards in nearly every product, system, or device. Embedded in everything from e-mail to social media to chat systems to Q&A sites such as Quora, this reward structure is omnipresent and not easy for us to control without going to extremes and without constant vigilance.

The Empty Vessel of Social Approval

When you post your first picture on Instagram, the application automatically contacts your friends who are already on Instagram and asks them to give you some "love." This is to encourage you to use the app more often and to get you hooked on social approval. It is a well-known product-design tactic in social networks and other consumer products. Both Twitter and Facebook encourage new users to immediately follow or connect with others they may already know in order to ensure that their feeds fill sufficiently to attract steady interest and to create a feedback loop of intermittent variable rewards. Sending some love seems rather innocuous, and the request is clearly not malicious in intent. But a little too much love can be bad for your soul when that love is empty and demand for it arises from a hedonic treadmill of empty accumulation rather than from real social relationships and personal recognition.

We all need and compete for social approval at some level, from our families, our friends, and our colleagues. Even if we intentionally try to avoid seeking it, the social-media software and hardware and their mass penetration via the Internet have led social competition to occupy considerable portions of our devices, our time, and our thoughts. Teens posting messages on the popular photo-sharing site Instagram worry acutely about how many likes and comments they will receive. To members of Instagram, followers are social currency. In Snapchat, teens compete to maintain "Snapstreaks"—consecutive days of mutual messaging—with friends. On Facebook, the number of likes on a post or the number of messages you get on your birthday becomes a measure of your personal self-worth. On Twitter, journalists and intellectuals compete for retweets and "hearts." On LinkedIn, we check to see who has viewed our profile, and the application provides us with weekly stats on the increase (as a percentage or an absolute number) in the number of people who have checked us out.

To be fair, some evidence exists that active participation in social networks leads people to feel more connected.[12] Facebook claims that chatting with friends and family, sharing pictures, and other positive interactions don't make people sad, although it concedes that negative comparisons can lead to less happiness.[13] Certain personality types, it appears, can better control the craving for constant likes

and approvals, and suffer less from the inevitable comparisons with those who are more popular.

But, in general, jealous comparisons kill joy, and technology has driven us to compare ourselves with others
on the most superficial of measures.[14] Furthermore, recent research on social-media use has found that it is the
comparisons, which are unavoidable in social media, that
contribute most to making users unhappy.[15] Teenagers appear to be particularly vulnerable to this; being excluded or
unloved on social media is one of the worst humiliations
a high-schooler can suffer.[16] Heavy social-media use has
been linked to unhappy relationships and higher divorce
rates.[17] That may follow from social media's encourage-
ment of social comparisons and self-objectification, which
tend to lower self-esteem, reduce mental health, and incul-
cate body shame.[18] Quitting social media has been linked
to marked increases in well-being.[19]

This behavior of seeking likes and approvals also relates
directly to intermittent variable rewards: the slot machine in
our pockets and on our tablets and laptops. Not knowing how
many likes you will get or when they will roll in, you check
your social-media accounts frequently. And limits on choice
and control compound the active promotion of destructive
behaviors to escalate users into borderline obsessiveness.

The Bottomless Well

It's 11 p.m. on a weeknight, and you reach the end of the first
episode of the latest season of *Stranger Things* on Netflix.
It's late, and you know you should go to sleep. You have to

be up in eight hours to go to work, and you need your rest. But before you can close the application, the next episode begins to play. Netflix has conveniently loaded that episode in the background, anticipating your desire to continue following the story. And then, almost against your will, you are watching the next episode even if you intended not to. *Oh well*, you figure, *I can make up sleep on the weekend.*

Along with the millions of others watching Netflix at that precise instant, you have just been sucked into the bottomless well of consumption. Netflix has teams of PhD data scientists who work to figure out how to get you to watch more movies. As you watch Netflix, they watch you, tracking your behavior in minute detail. They track when you pause, rewind, or fast-forward; the days of the week when you tend to watch; the times of day when you watch; where you watch (by zip code); what device you watch on; the content you watch; how long you pause for (and whether you return); whether you rate content; how often you search content; and how you browse and scroll—to name just a few parameters. Truly, they are watching you watching them!

So it's hardly surprising that Netflix figured out that starting the next episode without even asking you would entice you to consume far more content. They noticed that some users were binge-watching and decided that automatically activating the next episode might be a good feature. Netflix launched "Post-Play," as the feature is called, in 2012. Other video-hosting companies quickly followed suit. It got so bad that Apple built a feature into Safari that blocks auto-play videos on webpages and, in January 2018, Google made this a feature in its Chrome browser! So how much more do we consume when facing a bottomless pit of content? Real data on that aren't publicly available yet (although Netflix, YouTube, and Facebook certainly have them), but clues to the soaring amount of user time that Netflix, YouTube, and Facebook videos occupy are available in research and surveys. A 2017 report that surveyed 37,000 consumers found that Netflix binge-watching had become "the new normal," with 37% of binge-watchers actually partaking in their pastime at work![20]

Since Netflix launched the feature, every other major streaming video provider has taken advantage of the over-consumption that follows from automatic availability. Netflix, Hulu, YouTube, and HBO all have bottomless wells set up on their video applications. The lesson has not been lost on traditional online publications, either. Most media sites now offer suggested reading links at the ends of articles and in sidebars as well as highlighting "most popular," "most shared," and "most e-mailed" articles. Many of them, mirroring Facebook, Instagram, and Twitter, now have scrolling pages that cause each article to roll into the next without requiring a click. The goal is to boost consumption, at nearly any cost, even that of fostering a consumer's destructive behaviors. In effect, every digital company wants us to binge-watch everything, all the time. Our value to it has been reduced to the amount of time we spend in an application watching a video or playing a game.

This is hardly the first time that for-profit businesses have sought to induce addictive behavior. The soft-drink companies such as Coca-Cola, the tobacco companies, fast-food chains, and convenience stores such as 7–Eleven all focus on building repeatable habits for reliable long-term consumption of their products. They have done this largely without real concern for the impact on the user's or consumer's well-being. To those whose paramount concern is profit, such disregard makes perfect sense. Why would they suggest that those constantly tapping a screen to place more bets (literal or figurative) consider the impact of their actions on their families, their finances, and their health? But most of the large tech companies stake a claim not to operate in such a vacuum: they claim to be doing what they are doing in part to promote the betterment of humankind.

True, Coca-Cola, PepsiCo, and other companies peddling addictive products also have lofty mission statements. But society doesn't take their mission statements seriously, and neither do they truly have the potential to better humankind except in underwriting charitable efforts, as Coke will never announce that due to the link of sugary drinks with diabetes it will cease selling those drinks. In contrast, Facebook, Twitter, and other social-network tools do have a unique potential to effect positive change; witness the impact of Twitter carrying the message of the Arab Spring movement, and the use of Facebook as a means of recruiting subjects for trials of experimental drugs, a significantly cheaper technique than the traditional recruitment methods.[21]

Another key way in which the online and Internet giants differ from the others lies in their ubiquity—and therefore their power—in our lives. No one spends nine hours a day eating McDonald's or hanging out in 7–Elevens. You may carry a soda or a cup of coffee for several hours in a day, but you don't usually sleep next to it or take a swig of it in the middle of the night when you awake. You don't conveniently carry those experiences everywhere in your pocket and mount them on your dashboard. You don't totally freak out if you don't know where your soda is! The only exception we can think of is tobacco products. But even the most deeply addicted cigarette smoker can go for an hour or two without lighting up, whereas normal people who have a healthy relationship with online tools rarely go a full two hours during a working day without logging in, checking e-mail, or undertaking some form of social activity on line.

Equally troubling, recent research has associated binge-watching with sleep disorders.[22] Netflix CEO Reid Hastings stated, half in jest, that the company's primary competition is sleep, perhaps not realizing the truth in his words.[23] We return to the effects of media technology on our sleep in chapter 6.

So large technology companies' decisions to default us to the bottomless pit of content show that they may not have our own best interests in mind. To be fair, Facebook, Netflix, Hulu, and YouTube all allow users to turn off this auto-play feature (though apparently HBO Now does not). But wouldn't it be better for everyone if people could opt *into* the feature rather than encounter it and have to opt out? A simple Play Next Episode button works almost as well. And when we want to opt out of video auto-play on Facebook, arriving at the right setting takes a few not necessarily intuitive steps. This naturally discourages people from turning the feature off.

This may seem a paternalistic suggestion, but making such repetition an opt-in feature would give users a chance to make a more conscious decision before they are trained to expect auto-play. In pausing, we temporarily break a pattern, returning decision-making to our conscious minds and establishing a fresh opportunity to sidestep or counter our addictive behaviors. And the rarity with which tech and application vendors allow users to opt in rather than opt out—or even to pause—puts the lie to any claims of innocence. They know that far fewer users would consciously decide to drink repeatedly from the bottomless well; and profit maintenance takes precedence over user choice.

FOMO: The Gnawing Fear That We Are Missing Something Important

Fifty years ago, when we left the office or the job, we heard from our managers or employees only if there was a real emergency. Such communication would take the shape of a phone call. Today, notification inflation is part of every job. During an eight-hour workday, on average we check our e-mails nine times an hour.[24] We send texts to update our progress while we're in transit to the office or to let people know when we'll emerge from a meeting. Each of those notifications that we send in turn demands attention from its recipients. How many of those interruptions are necessary or even helpful? Probably fewer than 5 percent of them.

But these notifications are perceived as exceptionally valuable by the companies that make communication tools for work. For example, Slack was the fastest-growing business chat tool in 2017. It was worth more than $5 billion as of July 2017.[25] It looks a lot like nearly every other chat tool ever made, going back to IRC (Internet Relay Chat), but Slack uses numerous tricks to hook users and entice them to spend more time using the application.

In fact, the company is so convinced that constant notifications are a positive feature that its product designers resort to scare tactics should a user wish to turn them off. To ensure that Slack users buy into all this notification noise, Slack presents a stark warning when someone decides not to enable desktop notifications of Slack conversations: "Desktop notifications are currently disabled. We strongly recommend enabling them." Slack would probably counter that its users can turn on Do Not Disturb mode inside the app whenever they wish to concentrate, but that very argument implies that interruption as a default state is optimal. We beg to differ: interruption as a default state appears to be miserable, unproductive, and bad for our health.

On top of notification inflation, then, we have built a culture of FOMO: fear of missing out. We check our e-mail first thing in the morning to see what happened while we were sleeping. This fills our brain with unnecessary conversations during its otherwise most productive and creative time, the morning. (That would be a lesser problem if the average e-mail message were more useful.) Productivity gurus such as Tim Ferris and Cal Newport intentionally avoid answering e-mails or texts until after they have completed their most important tasks of the day. This makes perfect sense when we consider how often we check e-mail.

University of California Irvine researcher Gloria Mark and colleagues found that workers check e-mail an average of seventy-seven times a day—and that checking e-mail constantly tends to increase worker frustration and stress.[26] If we had checked our e-mails seventy-seven times on the days when we were writing this book, we would never have finished writing the book!

We keep people on as Facebook friends even though we don't really want to, because we are afraid that we might miss out on something that people in our high-school class are doing, saying, or experiencing. We refrain from unfollowing people on Twitter because they might notice and take offense. Yet we keep those same people unmuted in our feed just in case they post something interesting. We use tools such as Nuzzel to save time by giving us a newsfeed of everything that our friends are reading (or at least posting on Twitter), although this also means we have more to read and are less focused in our reading.

And we spend time on Facebook Messenger or Whats-App chatting about things that have little to do with our work, to see what we've missed out on around the virtual office watercooler. In the tech world, Slack is very popular. The neighborhood version of Slack is NextDoor. On NextDoor, neighbors connect in useful ways to share information and to chat, but they also spend many hours in vitriolic arguments over whether dogs should be leashed in the park or whether it's okay to light a wood-fired stove in the winter. NextDoor, too, strongly encourages accepting notifications.

In our use of every screen device, and on nearly every app and website, some kind of Do Not Disturb function exists: on our laptop or phone, there are options to control notifications; in the various applications themselves, there are notification options; and of course there is the on/off switch. But somehow we rarely use them. And many work environments have unspoken understandings that a worker must respond to any e-mail, text, or chat from a superior within a certain period or face unpleasant consequences. Being labelled "unresponsive" and "not a team player" is often the code phrasing for someone who prefers to focus on his or her work rather than constantly monitor e-mail and chat messages in order to respond to superiors or colleagues.

Forcing Us to Follow Their Agenda to Reach Our Agenda

Tristan Harris discusses how technology companies set our agendas for us by mirroring and magnifying brick-and-mortar stores' strategies for influencing shoppers. For example, grocery stores put the most popular products—milk and prescriptions—at the back of the store in order to draw shoppers past as many products as possible, and they put things such as produce and deli and dairy displays along the outer walls to encourage shoppers to circle the stores.

Tech companies place similar distractions in the way of their own customers. Facebook, for instance, routes people through the newsfeed before they can see an event they are interested in. Naturally, we get distracted by our newsfeed because there is always something new there. This results in further consumption of Facebook but slower progress toward our original goal (checking out an event).

Of course, whenever we use a free service, such as most of the social networks, bending users to the company's agenda to increase consumption of advertising is part of the price of entry. We all know and understand that. But maybe we would prefer a paid option with a direct-access option for key tasks and screens? Or maybe there's a better way to help us get directly to our intended destination. These are wishful and wistful questions. We have no illusions that such options will be forthcoming, as they would enable us to reduce our time in the application and redirect our attention for a few seconds or minutes per month, to the chagrin of shareholders and the cadres of mathematicians and computer scientists whose primary job it is to get us to click on ads. To be fair, Facebook announced in January 2018 that it would switch its algorithms to show in the newsfeed far more news from friends and family. But it remains unclear whether that also includes news articles or just personal updates. Alex, for one, has relatives with strong political views that oppose his own, and he would rather not see their postings of hyperbolic (and sometimes fake) news articles.

Tristan Harris dreams of a digital bill of rights that would mandate direct access: "Imagine a digital 'bill of rights' outlining design standards that forced the products used by billions of people to let them navigate directly to what they want without needing to go through intentionally placed distractions."[27] Though Harris has long received support in his quest from the mindfulness and productivity communities, he is now receiving support from unexpected quarters: hedge funds and employee-pension funds. Jana Partners, a multibillion-dollar activist hedge fund; and the California State Teachers' Retirement System (CalSTRS), one of the largest public employee pension funds in the U.S., sent a letter to Apple CEO Tim Cook asking the company to consider how iPhones and iPads impact the well-being of children.[28] A digital "bill of rights," however, remains wishful thinking in the United States. There is no clear movement to establish a bill of users' rights, even if it is a really wonderful idea on how to balance addictive product design with user choice and control. A handful of companies are trying to do this, and we'll talk about them in the final chapters as well as on the book's website (HackedHappiness.com).

By contrast, Europe has been steadily putting in place laws that are building people's online rights, step by step. The "Right to be Forgotten" gives people the right to ask online properties to remove results or information about them. Europe also mandates that any algorithm in use be explainable to humans. This sounds quixotic, but it just may have the salutary impact of forcing companies to consider that their users may have a right to understand how decisions about them are reached. Europe's data-privacy laws—Germany's being the most stringent—tend to lead U.S. laws as well in putting the burden of maintaining user privacy on the companies that collect the data and in placing real limits on the kinds of data they can collect and under what circumstances they can collect it.

Exploiting People's Inability to Forecast Time Spent

How often have you clicked on a notification to check what caused the red bubble to pop up and learned that someone has tagged you in a picture, only to look up thirty minutes later to realize you've been aimlessly browsing through the photos of online friends? This is the digital equivalent of a classic sales technique: "Can I ask for a minute of your time?" It relies on a deep feature—some might say defect—of our basic mental functions.

In almost every activity, humans underestimate how long it will take to complete a task or how likely we are to become distracted. Some services, such as Medium, try to help users manage expectations by posting reading time on each article. But, by and large, technology encourages us to dive into tasks large and small with the understanding that doing so will take just a moment, though in reality any task will absorb us for longer than we estimate. This bias is compounded by all the ways in which technology companies seek to distract us as we undertake a task, making it even harder to estimate how long it will take us to finish. Imagine how cool it would be to have an I Am in a Hurry button on a smartphone, or an application that would clear the way of distractions such as ads, inducements to click on other feed items, or other tricks deployed to drive higher engagement. (In fact, ad-blocking software is a de facto I Am in a Hurry button: the main reason users run ad-blocking is that ads slow down their online experience, according to the Internet Advertising Bureau.)[29]

The "time spent" bias is even more pronounced across the entire smartphone platform and our general use of technology. We radically underestimate the amount of time we spend with our devices—perhaps even by half. Participants in a small study in 2015 of twenty-three adults of ages from eighteen to thirty-three estimated that they spent roughly two and a half hours per day on their phones. In fact, they had on average spent more than five hours, in nearly eighty-five activities, per day.[30]

Exploiting the Availability Bias

You may remember as a child playing in your neighborhood without supervision, riding bikes, or going to the park, and then just walking or running home when it was dinnertime. Such freedom is a rarity for children today, because of parents' fears for their children's safety, doubtless affected by our endlessly scrolling newsfeeds. People tend to overestimate the likelihood that negative events will happen to them or their children. And the Internet is a giant machine for inflation of availability bias. Our newsfeeds fill our heads with horrible news from around the globe, conveniently curated (not necessarily for accuracy) by news-aggregation engines over which we all too often have little control.[31]

People are naturally attracted to catastrophic events,

and the Internet plays to this attraction by making it possible for us to read about child abuse, horrible crimes, and all manner of sick behavior or dangerous events transpiring not just locally (as was formerly the case) but anywhere in the world. In the news business, as the saying goes, "If it bleeds, it leads." On the web, this phenomenon leads to results that are far more serious. Even as the statistical likelihood of violent crime and of child abduction has (at least until 2015) steadily fallen in the United States,[32] parents have adopted ludicrous precautions, such as driving their children half a block to school or refusing to allow their children to ride their bikes in safe neighborhoods or explore the woods near their homes.[33]

This fear is related also to the *affect heuristic*, a feeling that comes over us momentarily as a psychological response to stimulus.[34] Psychologists believe that this human tendency explains why messages designed to activate emotions are more persuasive than other messages.[35] In other words, when we confront an emotive article in a newsfeed with a horrible headline, the article has a larger effect on our thinking, and on our belief that such horrible events are common, than if we read a drier, more clinical, or statistics-driven article presenting the same factual content.

Just a Few of the Many

These are just a few of the ways in which technology reduces our choices, offers us false choices, and persuades us to consume more than we need to. Dozens of books—from the Dale Carnegie classic *How to Win Friends and Influence People*, to the catalogue of thirty-three psychological tricks of modern advertisers, *Hidden Persuasion*, by Marc Andrews and Matthijs ven Leeuwen, to Natasha Dow Schüll's *Addiction by Design*, an exposé of the incredibly detailed deliberations of Las Vegas casinos—have covered the myriad methodologies, proven and unproven, that are used to influence our behavior.[36] In relation to modern technology, these efforts trace back most prominently to the writings and teachings of a respected and quietly famous Stanford University professor named B. J. Fogg.

◆ 2 ◆

The Origins of Technology Addiction

B. J. Fogg and His Disciples

In 2006 a young Stanford student named Mike Krieger took a class with Professor B. J. Fogg on persuasive technology. Fogg had his students build applications as a class project, and Krieger designed one that would encourage people to send happy photos to their friends living in places where the weather was bad, using a phone. (This was prior to the mass use of smartphones for sharing photos.) Krieger called the application "Send the Sunshine," and the idea of sharing photos stuck with him.[1] He later went on to found, with Kevin Systrom, a photo-sharing social network, Instagram. Facebook acquired Instagram for $1 billion in 2012. At the time, Instagram hadn't earned a dime in revenue. What Instagram had, as Krieger had learned from his time with Fogg, was the ability to entice and addict its users, some of whom spent hours a day scrolling through the images posted by others, and further hours planning the images they wanted to capture and post.

Krieger is Fogg's most prominent alumnus, but many others learned from him how to build habits in users, including those who took his now famous course focused on Facebook apps in 2007.[2] (Students in that course built Facebook apps that pulled in millions of users and

generated tens or hundreds of millions of interactions—
and they were not professional software designers.)

Another Fogg alumnus, Nir Eyal, wrote possibly the
seminal book for designers of modern technology products, *Hooked: How to Build Habit-Forming Products*. It's
required reading at many prominent start-up incubators
and accelerators, has more than a thousand reviews on
Amazon, and is a perennial bestseller.[3] Eyal founded two
start-ups in the early 2000s before going on to become an
investor, consultant, and paid speaker. He is unabashed
about his goal of helping companies create products that
induce "sticky" behaviors and draw the attention of consumers; each year Eyal hosts a Habit Summit in which he
shares his latest insights on how to hook consumers (and,
more recently, on the ethics of hooking consumers on software applications).[4] His book is a concise and practical application guide to much of Fogg's research in an area that
academics call "behavioral design."

Other Fogg alumni have gone on to found companies,
and many occupy influential product-design and product-
development roles at companies large and small in Silicon
Valley. It is not an overstatement to say that Fogg has god-
like status in technology circles.

Beyond Silicon Valley and technology-product-design
circles, Fogg holds far less business renown: although respected in academic circles, he has published no bestselling
books and does not go on the broad lecture circuit. He was
far less focused on building addictive behaviors than, say,
the makers of actual slot machines, whose efforts pre-date

the development of the app economy. But Fogg has had a disproportionate influence on Silicon Valley founders and on the ongoing quest of so many technology companies to design products that exploit the hardwired weaknesses of our slowly evolving brains.

Fogg's work is an intellectual continuation of the research pioneered by B. F. Skinner in the 1930s at Harvard University. Skinner's box riveted the world of psychology, but, curiously, behaviorism did not immediately catch fire. It was B. J. Fogg, in the mid 1990s, who showed its power. As a PhD student at Stanford, Fogg had been intrigued by the idea of using computers—then still an emerging technology—to shape and reward behaviors. His curiosity was driven by a simple observation: technologies, in particular interactive technologies such as phones, computers, and tablets, were becoming more common and occupying a greater part of our lives. Fogg devised experiments to test that thesis and found that technology was even more powerful than he had thought. Technology-mediated cues proved extremely influential upon human behavior and were far more malleable than levers and Skinner's physical cues had been. It turned out that "mental food pellets" were at least as potent.

Fogg made the logical leap and envisioned how technology could be used as a tool for good: to help busy people better manage their finances, learn new skills more quickly and easily, and adopt healthier behaviors. In a 1997 paper on his findings, Fogg wrote presciently, "Exactly when and where such persuasion is beneficial and ethical should be the topic of further research and debate."[5] As he continued his research, Fogg advocated the establishment of a new field of study that wove computer science and psychology together. He thought it should be called "captology" (from "computers as persuasive technologies").[6]

According to Fogg, you need three things to happen in order to influence a behavior. A person must want to perform an action or do something, must have the ability to do so, and must receive a trigger or a prompt. Without motivation, a trigger is useless and just serves as an annoyance to the person. If the desired action is too difficult or complicated, that desire serves only to frustrate and discourage the person. The triad seems incredibly obvious, as great explications often do.

Probably Fogg's most important contribution, however, was the insight that in order to change behavior, reducing the scope and complexity of the task is more effective than increasing the motivation. (Think of the endless video spool on Netflix and Facebook. The goal is to get people to watch more videos. To that end, product designers remove all friction—making it harder for users to stop watching the video than to continue watching it.) This is particularly effective when our brains emit the pleasure-inducing chemical dopamine in response to stimuli such as receiving likes, checking e-mails and finding that one has arrived, or being pinged in an online chat. So facilitating access coupled with pleasure inducement is a powerful combination for creating habits of consumption.

The field would come to be called "behavioral design," and Fogg has remained its intellectual leader. It has dovetailed with a burgeoning interest in behavioral economics and decision science led by scholars such as Daniel Kahneman, Amos Tversky, Dan Ariely, and the 2017 Nobel prizewinner in economics, Richard Thaler. The research of these and others has illuminated the previously undocumented "wetware biases" (cognitive biases) and flaws that help explain some of what makes humans so susceptible to technology-driven behavioral design even if it results in outcomes that make them unhappier.

Some of Fogg's research findings may seem head-smackingly obvious. Importantly, Fogg is not a clear villain in this story. Much of his early and continuing research explores not only core behavioral-design techniques but also how to use mobile technology to improve health outcomes and change bad habits. In recent interviews, Fogg has expressed misgivings that his findings are being used for profiteering and hoarding human attention in ways that are not good for people or society. But as with many other researchers (Einstein prominent among them), his work was easily enough incorporated into uses that far outstripped his initial ideas.

Fogg is hardly alone with his qualms. Even Eyal has expressed concern that technology consumption is out of control, that habits driven by the very insights he helped to popularize are coming back to haunt us. Joining Jobs and Gates, Eyal puts strict limits on how his children consume technology, for example. In this he joins the growing ranks of technology leaders expressing concern that we have not sufficiently examined the long-term human impacts of modern technology. Under the microscope, as you'll see in the next few chapters, things do not look very good.

• 3 •

Online Technology and Love

We are living in the Tinder era, when a swipe is a terminal judgment taken in an instant upon scant information—at a rate of hundreds per hour. This is entirely new to humans— the appearance of an apparently endless choice of potential partners. Were it that simple! This chapter looks at how the Internet has changed our views of love, of romance, and even of ourselves.

How Technology Undermines Our Love Lives

Since the first civilizations, and across all cultures, humans have told stories about love. From Paris and Helen, to Romeo and Juliet, to Bonnie and Clyde, to Brangelina, lovers have captivated our imaginations, and love stories have become part of our cultural fabric. Very few of us can live happily without the love of others. The love of children and partner, of parents, and of friends: all contribute mightily to the richness of our existence.

In many ways, past technological revolutions have affected how we love. Universal schooling and the popularization of letter-writing made love letters a common vehicle of expression. Later, the camera allowed soldiers to exchange pictures with their wives and families and girlfriends. The telephone connected distant lovers and friends over twisted strands of copper wire.

These technologies surmounted distance to help us maintain our relationships through "social snacking," a taste of the experience of actually sharing time together. The Internet and smartphones have, however, made these tastes more lifelike and immediate, for better and for worse: for better because we are more connected and closer than we ever were; for worse because, since the arrival of the Internet and smartphones, we have seen pornography use skyrocket, with clear negative effects on relationships and fidelity. Divorce rates, which rose rapidly during the early days of the Internet, continue to rise in people over age fifty; marriage rates in the United States and the developed world are falling steadily; and key measures of intimacy in America, such as frequency of sexual intercourse, have declined.[1,2,3]

Other contributors to these declines undoubtedly exist, but the way we use the new technology seems to have contributed. Despite all the talk of our being in a post-marriage society, married couples continue on balance to be happier and more stable than single people are, according to much of the research on the topic. And according to research by John Helliwell and Shawn Grover published by the National Bureau of Economic Research, that appears to be a causal relationship rather than a mere correlation.[4] Of course, lower rates of marriage correlate with higher poverty rates in general, particularly in the United States, where social safety nets are weaker than in other developed countries. But living in a two-parent household is a core indicator of well-being among children.[5,6]

Online technology has done a lot of good. Evidence is emerging that online dating has led to a marked increase in interracial dating and has enabled people to connect with others whom they would otherwise not have met.[7] For those living in remote locations with small populations, online dating may, by expanding their pools of possibilities, be the best way to find a potential mate. And our ability to communicate with loved ones using technology has brought great benefits to business travelers, who can chat with their spouses while on the road and share pictures with their children and extended families.

In this section, we discuss how modern technology has affected our love lives, focusing on some of the more obvious topics—online dating, pornography, divorce—and connecting those topics to broader trends that suggest a diminution in the experience of love and romance in our lives. As they do in other parts of our lives, always-on devices, along with the attention economy and the explicit design of technology products to maximize consumption, produce tedium, frustration, and dissatisfaction, both directly and indirectly.

We see this occurring through our acceptance of ways of seeking love that seem practical and helpful but may in fact cheapen and trivialize what might otherwise be a less comfortable but ultimately transcendent experience. And some of the effects imply that moving love on line makes people's experience of seeking an ideal match more unpleasant, more time-consuming, and more stressful than otherwise, with less fulfilling results. Compounding that, online dating promotes expectations to which our love lives, fraught with comparisons, never seem to measure up.

Online Dating and the Endless Search for the Perfect Mate (or Just for Disposable Love)

Fifteen percent of Americans have used a website or app to look for a romantic partner.[8] The online dating industry now generates billions of dollars in revenue. It has also broken down barriers to finding potential mates as no previous technology has. We can now easily find and date people from different neighborhoods, in different towns, in foreign cities while on business trips, or in social groups that we might not otherwise regularly encounter in our daily lives. We can filter online referrals to potential mates on virtually any parameter we choose to zero in on, talking to only those who have the characteristics and background that we want in a partner.

Online dating remains new enough to raise many questions about its effects. How have online dating and dating applications affected human fulfillment? Has the endless catalogue of swipes and potential mates led us to better marriages and greater fulfillment? Or has it led us astray and left us feeling sadder and lonelier? Has it led to fidelity or to divorce? The answers to these questions are complicated. Over all, online dating appears to have diminished our ability to find romantic satisfaction. More troubling is that online dating may be changing the way we view romance and how we assess and value other human beings, making our views and values more superficial, overly influenced by physical appearance.

The experience of our good friend Alice (not her real name) is a case in point. A divorced mother with college-age children, Alice has had an amazing career, rising to high levels at prestigious companies as a technology executive. Outside of work, she is creative and an avid hiker. She paints, makes pickles, and helps her many artist friends turn their work into decent businesses. She is funny, smart, and full of energy. We have known Alice for four years—and in that time, she has always struggled to find good dating partners.

This lack of dates is not for want of trying. Alice has used several online dating sites and gone on a handful of dates, but she found the entire experience unsatisfying. She considered Tinder but found it superficial and somewhat obnoxious. Even with more-thoughtful dating sites such as OkCupid, she felt that the dates never lived up to the online conversations or the expectations that the profile pictures raised. That's perhaps unsurprising, considering that the majority of people are at least somewhat dishonest in their profiles.[9] She grew so dissatisfied that she began organizing parties to which her single friends brought their own single friends to meet and greet—not unlike an old-fashioned dance or social.

Alice's disappointment with online dating seems to be

common. In response to a survey by Consumer Reports, the most active group of people using online dating sites found the experience incredibly frustrating: those who were most active gave online dating sites the lowest ratings of any service sites across all industries.[10]

To be sure, 44 percent of respondents to the survey said that online dating had resulted in a long-term relationship. This is certainly a positive result, and there are many success stories. One large study, too, suggests that marriages coming from online dating are happier.[11] These positive outcomes may represent a selection bias: more-systematic people may tend to use online dating for successful outcomes precisely because they are systematic in their approaches and have a specific strategy (a wonderful book about this is *Data: A Love Story*, by Amy Webb) or because a better attitude toward dating led them to use a certain dating site in a more thoughtful way.[12] If so, those people probably would have ended up with a better marriage result anyway, even if online dating had not existed. (The study was funded by the online dating site eHarmony.)

Also, a fundamental change has come about through online dating. In the past, online dating was an intentional act. People logged on to a dating website to browse, chat, and look for partners. The website was separate from other online activity and was not set up by default to induce addictive behavior, though it played the usual tricks to foster engagement: e-mail alerts, phone applications, and so forth.

Then along came Tinder, with swiping and other clever user-interface tricks that foster the actions of rating, comparing, and selecting potential mates. This changed online dating in a key way. It became an omnipresent activity, a sport almost—swipe left, swipe right—that Tinder users could play in bars, in elevators, on the subway: anywhere they happened to be. Tinder's innovation made online dating more addictive and also more comparative in an unhealthy way. Before, it took effort and focus. Tinder and its ilk made it mindless and ceaseless: the dating-application companies essentially used their simple navigation schemes to transform online dating into a form of endless mate shopping. On the one hand, we can admire its elegance. On the other hand, we see the result. The research discussed in this chapter shows that Tinder is bad for self-esteem and generally bad for the psyches of its users.

And the proof is really in what people do, rather than in what they say. If online dating were a panacea for romance, we probably would have seen soaring rates of romantic engagement. And we have not. To the contrary, in the era of modern technology, we have gone in reverse: more of us live alone; we marry later; we have fewer partners; and we spend less time in social groups and outside our homes.[13] That said, it is critical to remember that technology is neither good nor bad except as it is used or abused.

And our cultural expectations may play a significant role in establishing that trend. Vivek has two Indian American relatives who approached online dating sites not for dates but with a view to matrimony and building a family. They used Indian dating sites and found lifelong partners and much happiness through those relationships. There are also likely to be cultural differences between small towns and large cities, between rural and urban areas, and even between different members of the same religion in terms of how they use online dating sites and their purposes in accessing them. For this reason, we try to focus on aggregate numbers.

Examining the True Effects of Online Dating

The effects of online dating and dating apps on happiness are complex. On the one hand, online dating exposes people to a far wider set of options and allows filtering by criteria of the user's choosing. On the other hand, the paradox of choice affects many by making a decision difficult—and when they do make a decision, they tend to be less happy with it.[14] This may occur because that style of online dating promotes a mentality that views people and relationships as commodities to shop for.[15]

This focused and highly regimented shopping for mates, which occurs even on an app as apparently unstructured as Tinder, also precludes the wonders of serendipity. In a column in the *New York Times*, author Maris Kreizman tells how, after failing to meet a nice match through algorithm-driven online dating sites, she found happiness with a love partner who was too young, in the wrong profession, and living in the wrong neighborhood.[16] Naturally, they met the old-fashioned way, in a bar through friends of friends:

> "In walked a friend of a friend who I sort of knew from the internet but who I'd never met in real life. He is six years

younger than I am (way too young for me) and he lived in Harlem (that's a $40 cab fare from my home in Brooklyn) and he's a writer/comedian (warning flags coming at me from every direction). But we talked and he charmed me. He was online dating, too, but I never would've found him on an app. He wasn't on my metaphorical vision board, but he fit into my real life in ways I never could've imagined. He's my husband now."

Aside from eliminating serendipity, online dating in its current format promotes a winner-take-all effect, wherein everyone seeks to date the most attractive people.[17] This eliminates selection of mates by other variables that may be more predictive of compatibility, leading to frustration all around. There is no way to set up an online filter for people whom you find funny or fascinating or endearing, because those qualities are subjective and because they emerge in how people interact in a relationship. For these and other reasons, we believe that online dating is likely to be making it hard for people to find fulfillment in the meaningful, lasting relationships that contribute most to their lives.

Since the advent of Match.com in 1995, online dating has exploded into a diverse set of applications and services, from "science-backed" dating sites such as OkCupid, Chemistry, and eHarmony; to affinity-group sites such as JDate (Jewish) and Grindr (gay men); to swipe apps (Tinder); to apps that give more control to women (Bumble). Advocates of online dating claim that it improves choice, makes finding partners easier, and improves our ability to select partners on the basis of what we perceive to be compatible characteristics. A significant and growing body of science implies otherwise.

Through a comprehensive survey, psychologist Eli J. Finkel and colleagues found that online dating "does not always improve romantic outcomes." This occurs for many reasons, one being that online dating is reductive, narrowing social interactions to two dimensions.[18] The experience of meeting someone on line fails to capture the essence of social interaction. Online dating, Finkel and colleagues write, is evaluative and creates a mindset of assessment rather than of engagement.

This mindset leads online daters to view potential partners as just another online commodity, undermining their willingness to commit because there is always a better option out there, even if they don't know about it at that instant. Indeed, it is probable that those who are most successful at online dating attack it with ruthless efficiency, playing a high-volume game and quickly casting aside anyone who doesn't (either on line or once dates commence in real life) generate an immediate spark of interest.

This rating culture and commoditization mindset may also lead to diminished appreciation of people before we even meet them. Scientists are coming to believe that physical attraction is not fixed. We change what we think about people's attractiveness based on our interaction with them. Funny people or clever people or extremely empathetic people may become more attractive to us after we talk with them or spend time with them.

Kansas University researchers documented this effect,

calling it "the Tinder trap."[19] In a lab setting, they showed subjects pictures of potential mates and asked them to rate their attractiveness. The researchers then introduced some of the subjects face to face to the people they had rated. The scientists found, curiously, that potential partners they had rated as less attractive or moderately attractive were far more likely to get increased ratings after a face-to-face meeting than were potential partners they had rated as attractive. So evaluating a potential partner solely on visual attractiveness is a poor predictor of what you will think of that person once you meet in real life.

Perhaps most importantly, rating people's attractiveness prior to meeting them tends to diminish raters' evaluations of them afterward, "probably because the rater is comparing their conversation partner to all the other potential partners they saw on line."[20] In other words, the apparently endless choice that online dating offers may cheapen and undermine our perceptions of people in real life.

The rapid-fire judgments inherent to online dating may also encourage our worst instincts. In a review of research about online dating in 2012, Eli Finkel and colleagues found little evidence that online dating was a net positive.[21] Their review was performed prior to Tinder's rise in popularity, but it foreshadows Tinder-like behavior. And it frames behaviors encouraged by online dating as basically antithetical to the outcome of identifying and loving a partner.

For example, the variety of choices in online dating can adversely affect our judgement in insidious ways. Evaluating choices side by side tends to encourage daters to emphasize factors and characteristics that are unlikely to determine compatibility. Whether someone is taller, or has blond hair or red hair, is highly unlikely to reflect compatibility over time; far less so than more-innate traits such as empathy, intelligence, or humor, or that elusive quality that emerges only in face-to-face interaction, "chemistry." Particularly useless in this regard are superficial physical traits that tend to be overemphasized due to reliance on photos as the primary basis upon which to choose a date. Finkel and colleagues call it "relationshopping," saying, "Much like hunting for size 8 leather shoes on Zappos .com, online daters seek partners by searching through profiles using attributes such as income and hair color, as opposed to arguably more important factors, such as sense of humor or rapport . . . We are bad at predicting what we will find attractive in real life."[21]

The snapshot-selection process not only leads many online daters to fudge their appearance but also tends to dwarf other considerations. OkCupid, which was founded by Harvard math majors, went so far as to measure the impact of what people wrote in their profiles in comparison with the pictures they posted. They found that what people wrote about themselves mattered little in determining perceptions of their attractiveness.[23] And, ironically, exposure to so many choices lowers satisfaction with the mate chosen.[24]

The multitude of choices also results in scattered, unfocused communication, which is likely to contribute to the transactional nature of "relationshopping." In a metadata analysis of four hundred thousand online-dating interactions, researchers found that the vast majority of the conversations are short, under twenty messages in length, and last an average of fifteen days.[25] In a Hobbesian twist, only 1.4 percent of the conversations result in an exchange of phone numbers. Imagine, then, going to a bar and introducing yourself to ninety people in rapid succession, of whom only a few are willing to talk to you, and leaving with one phone number. That experience sounds exhausting and miserable rather than liberating! This has led to "dating-app fatigue," with some women saying they spend ten to fifteen hours a week managing their online-dating lives. As Julie Beck writes in *The Atlantic*, dating applications "facilitate our culture's worst impulses for efficiency in the arena where we most need to resist those impulses. . . . Efficient dating is, in many ways, at odds with effective dating."[26] Or, put another way, focusing on quantity over quality is exhausting. Rather than searching for the perfect mate or a needle in a haystack, people might be better served by just trying to get to know each other better and letting their intrinsic attractions emerge.

More concerning is that some online-dating applications have been linked with low self-esteem. In a survey of Tinder users and nonusers, those who used the swiping app recorded lower levels of self-worth and, along with other negative impressions, said that they were less satisfied with their own faces' appearance.[27] Curiously, this effect was stronger in male users.

Now we move from how technology is affecting our

dating lives to how it is affecting our sex lives—through pornography.

Pornography

For the purposes of this section, we leave aside all moral judgments about viewing pornography. Rather, we focus on the impacts of its consumption on our lives. It is an uncomfortable subject, but online technology has taken pornography out of the closet and into the mainstream. Pornhub, the largest pornography site on the Internet, is ranked as the thirty-eighth most trafficked web property in the world, according to Amazon's Alexa traffic-ranking service. It tallied 2.66 billion sessions in November 2017, more than double the number of monthly sessions eight years earlier. Estimates of the percentage of worldwide web traffic devoted to pornography range from 5 percent to 20 percent (weighted by its heavy video content).

According to Pornhub's 2017 *Year in Porn Review* report, visitors watched four billion hours of pornography over the course of 28.5 billion site visits to Pornhub alone, not counting visits to the thousands of other online sites and blogs that publish pornography.[28] Visitors performed 24.7 billion searches on the site (about eight hundred per second, the number of hamburgers that McDonald's sells per second). No one doubts any longer that a fairly large percentage of the population consumes online porn. International studies estimate that 50 to 99 percent of men and 30 to 86 percent of women consume porn, the vast majority on line.[29] The average session on Pornhub was seven minutes in length as of November 2017.

Though some psychologists believe that porn consumption is innocuous and even associated with reductions in reported sexual assaults, a substantial volume of research indicates that porn is not so benign. In a survey of 1,500 people, researchers found that people who viewed porn even once a month expressed lower degrees of sexual satisfaction than those who didn't, with "disproportionately larger decrements in satisfaction" in those who consumed it more often.[30] In another study, researchers found that couples of which neither member used porn reported more relationship satisfaction than did couples of which one person used porn. Individual users, the study found, reported "significantly less intimacy and commitment in their relationship than non-users and shared users."[31] Other research associates the use of pornography with a higher likelihood of cheating on spouses.[32] Researchers have found that regular, heavy porn use may physically shrink parts of our brains.[33] Roughly one-third of married women view surreptitious or unapproved use of pornography by their spouses as a form of infidelity.[34]

This is hardly a settled topic, and a debate continues to rage over whether online porn is actually physically and psychologically addictive. One researcher found that treatment of people who believed they were porn addicts with Naltrexone, a drug used to treat drug addictions, significantly lessened their time spent on line consuming pornography.[35] Some counter-evidence indicates that joint porn use within a relationship increases females' reports of sexual intimacy and quality.[36] And porn watchers were more likely to be having sex than non–porn watchers, according to the most recent survey.[37] But having sex and making love are not equivalent, and most research suggests that porn has a negative impact on us, on our self-perceptions, and on our love lives.

Psychologists and researchers are concerned that sexuality is becoming divorced from intimacy—a trend that could accelerate as improvements in pornography technology make it a better and better replacement for intimate sex with people we love.[38] If porn in fact becomes more attractive than the real thing—more convenient, more enjoyable, and sufficiently realistic—and becomes more widely consumed, numerous other problems could result. Sex is effective not just for procreation; it has multiple beneficial emotional and physical effects on us. Having good sex with someone we love increases our happiness and well-being and may even increase how long we live![39] Sex with somebody we love confers a range of important health benefits, and it's possible that the intimacy associated with sex is more important than the sex itself.[40] Replacing live sex with pornography could have many unforeseen unfortunate consequences. Yet we may be heading down that path without asking what it may do to our relationships and to the meaning of being human.

· 4 ·

Online Technology and Work

Before the Internet and smartphones, we left our work at the office. We typed reports on typewriters (or on word processors and PCs). Calling meetings was complicated. Today, e-mail and chat connect us around the clock to colleagues around the world. We share information quickly and easily—and are often interrupted. We must process and filter far more information. As we show in this chapter, the changes that online technology has wrought in recent years may be impeding our work, reducing our productiveness, and taking its toll on our well-being.

The Hard Thing about Deep Work

As Alex sat down to work on this chapter for the first time, he shut off his Internet access and settled in to read the research he and his research assistant, Sachin Maini, had collected on the impact of technology on the workplace. About fifteen minutes into reading on his computer screen off line, Alex clicked on a URL because he wanted to read a related article. He turned his Internet access back on, went to that article, and found that he would need to search the web for a PDF version.

This meant a side trip to Google Scholar and an additional search. The search turned up more than a dozen related articles, many of which appeared to be interesting. Alex filtered the responses by year and soon realized that a lot of the Google Scholar results pointed to low-quality research. But they gave him ideas for other avenues of inquiry, so he entered some new search terms into Google Scholar and pulled out new articles. One of the new articles looked particularly promising, so he e-mailed it to Sachin and Vivek. Twenty minutes had passed since he had diverted his attention to the Internet. Alex had lost his flow but learned a few things along the way.

All of that would have been fine, except that Alex was on a tight deadline to finish this book. He again turned off Internet access to start over. But he also realized that writing a book in twenty-minute spurts of concentration interrupted by Internet research and e-mails was a poor strategy. As a university student, he had been able to write for two to three hours at a stretch without interruption, maintaining a more or less consistent train of thought. As a child, he had often read for five hours straight with scant rest. For his part, Vivek had whiled away entire afternoons coding without feeling the need to do anything else. Concentrating was just something you did; you didn't have to engineer a specialized existence in order to prevent distraction.

What had happened since those days? What had made it so much harder to get long stretches when you could concentrate and get tasks done? You already know the answer to this question. Technology has made it a snap to research books, find information, and share that information. Researching a book like this in the 1980s would have required weeks in a university library looking at microfiche and dead-tree journal articles. Now, such research can happen much faster on line, and even on mobile phones, and the articles can be easily shared with research assistants.

What we now have trouble doing, however, is staying focused. The distraction and the difficulty in focusing are so great that despite the far easier access to much of the world's knowledge, the fractured path from A to B that results from the addictive and interruptive temptations embedded in technology may be making us less productive than we were before.

There's no doubt that technology can make work more productive in key ways. E-mail and chat have made sharing documents and collaborating and conducting asynchronous communication across time zones and teams easier—at least in principle. Technology enhances our access to knowledge and information, making many tasks far more efficient. Think of the plight of a repair worker in the field sifting through paper manuals in the 1970s. Today, workers can receive the latest maintenance updates on a phone or a tablet and quickly find precisely the information they need. But they may feel compelled to respond to unimportant work e-mails during that same work task.

A growing body of evidence suggests that the darker side of technology use—distraction, interruption, and resultant lags in resuming concentration—is likely to frustrate and stress us and impede deep, sustained, thoughtful work, the kind that continues to be the most valuable to organizations and businesses as well as the most satisfying to workers. (*Alex dives in to check his e-mail. That takes just fifteen seconds, but it incurs a huge cost: at least five minutes more for him to mentally settle back into his original task!*) Nearly all the evidence points to our having become scattered, interrupted, and detained in meetings to a far greater degree than in the past. (*Alex shuts off Internet access but then turns it back on to leave a comment in a document for his research assistant.*)

Our personal experiences in writing this book were more distracted than we would have liked. This distraction is, we believe, increasingly representative of most people's experiences of technology and work. Pervasive social media, mobile applications, and other technologies use powerful psychological tools to hijack our attention. And companies seeking to win the war for our attention at work are using the same arsenal.

Worse, our employers themselves want more of our attention in ways that further dilute our focus, in ways formerly reserved for our personal lives. Employees are commonly expected to share company news on social media—and on the growing list of tools designed to coordinate and amplify marketing campaigns via individual employees' messages. In other words, our employers are actively encouraging us to repeatedly interrupt our own work!

Whereas pressure to attend to social media in our personal lives arises through manipulation of how we perceive our personal relationships, on the job there are strong corporate pressures to respond and participate. In survey after survey, workers say they feel compelled at work to check e-mail, text, and chat in order to respond quickly to any messages they receive or issues that crop up. The companies and organizations that foster this mindset give almost no consideration as to whether our always-on cycle for response and participation is ultimately productive and enhances both our personal satisfaction and the efficacy of the organization.

We believe that the adverse effects of new technology in our work lives have begun to outweigh the beneficial ones. Communications technologies and workplace norms put us in a constant state of alert and are eating into our time to focus on individual deep work. More concerningly, organizations are increasingly using technology to monitor and surveil us at work. The argument for surveillance is that employees will be more productive if they can get real-time feedback. The reality is that this type of surveillance and feedback—which comes in the guise of gamification— is simply another cause of distraction, frustration, and bitterness.

In the past decade, productivity increases in the U.S. economy have stalled. Economists have posited explanations—primarily, that it takes a long time for the impact of big technology shifts to show up in total factor productivity numbers, the gold standard for measuring productivity. This is not unusual. In past technology shifts, we saw a similar period of stagnation before a big leap. But other researchers propose a different explanation for continuing productivity stagnation: that, as it is currently used, technology at work is detracting from productivity.

The Productivity Paradox and Polyani's Paradox

Economists have long struggled to demonstrate a clear relationship between the adoption of the latest wave of technology—computers and the Internet—and improvements in productivity. Since the 1960s, they have closely tracked the impact of computers and information technology on how we work and on how efficient we are. And in the earlier periods of technology adoption, economists tracked rapid increases in efficiency, although those were largely concentrated in the technology companies themselves.

But a fundamental problem of productivity measures is that they were designed to measure production of physical things rather than of services. Production of physical things has become increasingly automated as industries prioritize efficiency. As a result, employment growth in the more advanced economies has been in knowledge and service sectors. Measuring output in this new economy is extraordinarily difficult, and economists continue to struggle with it. Measuring outputs of goods such as software is equally challenging. And productivity measurements don't accurately account for the benefits of price declines. For example, in 2005 we paid several hundred dollars for a dashboard Garmin GPS unit. Today, GPS is available through free applications on our smartphones; we pay nothing for the same functionality. The cost reduction in real dollars is probably close to $1,000. Applications such as WhatsApp and Skype, which allow us to talk for free, and even the calculators that are a basic feature of every computer and phone, are also illustrations of things that used to cost money and now cost nothing (or, after you buy the phone and pay your phone bill, next to nothing). There is no easy way to factor such improvements into current statistical analyses.

Measuring productivity improvements resulting from the introduction of new technology is as difficult as measuring productivity in the new service economy. The effects are elusive and complicated.

We intuitively know, for instance, that e-mail is exceptionally helpful for distributing documents. But we may not accurately account for the detrimental effects of spending too much time checking for incoming e-mail messages— or of relying on e-mail rather than on face-to-face conversations. (As we'll detail in this chapter, research has shown that live conversations are far more likely to elicit a meaningful response from people we don't know than e-mail is.)

Similarly, even if we are aware of them, researchers cannot easily measure the negative effects on productivity of the scarcity of opportunity to engage in "deep work"—a term popularized by computer scientist Cal Newport to describe work that requires extended thought and concentration.[1] Newport makes a strong case that such work is behind most of the great intellectual advances of our time and that the ability to engineer a life that supports deep work is a critical component of success. Newport builds on mountains of research concerning "flow," a psychological state popularized by psychologist Mihaly Csikszentmihalyi. Ability to achieve a "flow" state is widely accepted as the core underpinning of how the world's greatest athletes and musicians attain and maintain greatness.

In what has become known as "Polyani's paradox," Michael Polyani hypothesizes that we are less susceptible to automation and robotics than we think we are, because most of the jobs we do today have subtle nuances impossible to reduce to explicit instructions. Folding laundry, for example, is a task difficult for robots because it entails a great many skills that, though we take them for granted, robots don't possess. Let's say you tell a robot to gently tug on the corners of a sheet to stretch it out before folding. What does "gently" mean, exactly? And what if the sheet is a fitted sheet, without obvious corners? What is the difference between a round corner and a square corner? And can you describe the difference between a T-shirt and a sock in a jumbled pile of laundry? Until it can be made precise, no machine can successfully interpret the simple instruction "Please fold my laundry."

The tasks we perform every day at work, too, are far more nuanced than we may appreciate. Sending out a new report by e-mail to get feedback may actually result in a lot less feedback than handing each individual a copy of the report, looking them in the eye, and asking them to please read it and give you feedback on it. In imposing an intermediary, we may think we are increasing efficiency and improving productivity, but we are also bumping into

Polyani's paradox: it's far harder than we may realize to capture the full importance and scope of work activities. Interpersonal interactions have long been part of the process, and removing them wholesale inevitably has unintended consequences.

We build on the themes that Nicholas Carr explored in his book *The Shallows*, about the effect of the Internet on our brains: work activity is growing shallower, more fractured, more time-consuming, and less rewarding; productivity growth slows; and workplace uncertainty increases.[2] According to Stanford University professor Jeffrey Pfeffer in his book *Dying for a Paycheck*, technology companies (which, unsurprisingly, tend to be technology intensive) are increasingly unhealthy places in which to work.[3]

Technology has totally obliterated the border between work and other parts of our lives, leaving us no safe zone. On the one hand, some argue that this is a good thing, as the ability that technology fosters to work from anywhere on any device gives us incredible flexibility. It lets us run errands during short breaks on work-from-home days and allows us to keep working, without coming into the office, even if we are sick. Some knowledge workers are able to take "workcations," in which they visit a foreign country and work a regular day but after hours and on weekends experience the culture and surrounds of the place they're visiting.

On the other hand, this same facility for remote work reinforces unhealthy expectations that we should constantly be responsive to work demands. As well, it increases fragmentation of our work efforts by further segmenting our days into smaller discrete stretches of work, impeding deep work, the kind that is the most satisfying and productive. And the technology makes it very easy for us to multitask, which too easily reduces productivity.

Some work organizations have tried to mitigate this effect by aggressively enforcing absence from work during vacations and by experimenting with requirements that employees leave their smartphones at work when they go home. For example, German automobile manufacturer Daimler AG allows employees to set their e-mail accounts to "delete" while on vacation. Doing so automatically sends a reply that reroutes message senders to the person responsible for covering for the vacationer.[4] Notably, Google's human resources (HR) department strongly encourages employees to take vacations because it has found that rested employees are more productive and happier. Other tech companies are taking an even more radical approach by paying employees thousands of dollars during their vacations to refrain from answering e-mails or otherwise engaging in work tasks or discussions.[5] That companies are paying employees to ignore digital communications—an escape that employees should be able to relish during their time off—speaks volumes.

All told, the evidence strongly indicates that the turbulence, psychological burden, and constant interruptions resulting from modern workplace communications technology are disrupting our work and harming our well-being.

Information Overload, Communications Overload, and Metcalfe's Law

It is by now a truism that we live in an era of information overload. This arises from easy access to articles, songs, movies, and podcasts courtesy of the Internet. Many of us could spend fifty hours a week reading the latest news in our feeds and still not come close to keeping up. As information has become more accessible and its volume has burgeoned, we are increasingly required to filter its inflow—a task to which our evolutionary development has poorly suited us.

The rapidly evolving technological landscape and deluge of new information we are exposed to every day forces us to struggle to keep abreast of developments in our industry that are relevant to our professional lives. Filtering out noise, in the form of irrelevant and useless information, becomes ever more difficult, and the "signal," the truly relevant and useful information, remains increasingly elusive.

Exacerbating this conflict is our workplace technology, which now fights for our attention and competes in the attention economy nearly as aggressively as advertiser-driven models such as social networks and search engines. E-mail, text, and text-based chat now swamp voice in terms of time we spend on them per week.[6] This is unfortunate, because the newer forms of communication, though easier to digest and deal with, tend to be far less information-rich than voice conversations or in-person meetings. In a phone call, the tone of a conversation is easy to discern. In an e-mail, such nuances remain obscure. Additionally, flowing conversations, on the phone or in person, can yield insights or allow for the type of free association that enables creativity. In contrast, e-mail creates stilted, transactional interactions with little tendency to promote the free association and logical leaps that can lead to better ideas.

Metcalfe's law states that the value of a communications network increases with the square of the number of users. That's why Facebook as a network is so incredibly valuable: it connects two billion people, which is more than any nation-state or any religion except for Christianity does. The darker side of this relationship emerges as the costs of communication approaches zero. Organizational expert Michael Mankins posits that "as the cost of communications decreases, the number of interactions increases exponentially, as does the time required to process them."[7]

Today, communications are virtually free. Yes, we pay for network usage, and we pay for our cell-phone and broadband access. But we rarely perceive there to be a cost associated with sending an e-mail or a text or chat message, because in the developed world there is virtually no added cost in most cases. Free Wi-Fi is becoming more common, particularly in urban areas, and sending a business e-mail costs nothing. That has spurred the explosive growth in interactions predicted by Metcalfe's law.

The Radicati Group projected in February 2017 that by the end of 2021 the volume of business and consumer e-mails sent and received would reach 319.6 billion messages *per day* (and reached 269 billion by the end of 2017).[8]

WhatsApp, which employees use for both business and personal communications around the world, is relaying fifty-five billion messages a day free of charge, a total that exceeds the entire global volume of traditional SMS texts.[9] Over the course of a few years, Slack has grown to become a popular workplace chat and team-collaboration application. Many others, such as Atlassian's Stride (formerly HipChat), offer similar capabilities. Today, chat is, in many workplaces, deeply embedded as a third online communications mode after e-mail and text.

In Silicon Valley and the technology realm, Slack is now the dominant work chat application. As of October 2017, it was used by more than nine million people, in fifty-five thousand companies. Many of its users leave Slack running all day on their phones, receiving Slack notifications by e-mail or by SMS alerts on their phones. And, following the pattern of consumer technology's translation into work-productivity tools, the growth of chat applications at work reflects the rapid growth of consumer chat accounts. In 2017, five billion people were using chat tools globally, according to the Radicati Group.[10]

Naturally, there is a hidden cost to all that frictionless loquacity. Even as Slack touts its ability to radically improve productivity, numerous critics have decried uncontrolled Slack (and chat) use at work as doing the exact opposite: creating a giant time sink that also magnifies FOMO and other human information-processing flaws such as recency bias: the tendency to believe that what has been happening lately will continue happening.

This shift in how we interact at work is compromising work quality and productivity in profound ways that also detract from worker happiness. The productivity cost is much greater than the time spent looking at e-mail or checking Slack: employees are less "present" in meetings, less able to focus on tasks, and less able to devote uninterrupted time to critical tasks.

Evolutionary selective pressures did not favor brains with an ability to maintain perfect concentration while flipping back and forth between tasks. UC Irvine's Gloria Mark constructed careful studies that measure workers in real work situations or near-realistic work simulations, such as reading from a document and following instructions on sending e-mails to colleagues.[11] Mark found that interruptions can increase the total time necessary for completing a task, often significantly, and that it usually takes twenty-three minutes to return to a task after an interruption. At work, she found, these workers were switching tasks every ten minutes or so.

The task switches that do the most damage are those requiring a worker to switch contexts: from one project to another, or from one topic to another. Interruptions that remain within the same context are less disruptive. So, for example, an e-mail interruption about building slides for a report is not so bad if the worker is at that moment working on the report as it is if the worker is working on something else, such as writing performance reviews or reviewing a white paper.

Mark found that interruption even of tasks requiring less mental energy—such as sending an e-mail—stresses those interrupted, even if they can make up the time, possibly because they are left with less time in which to complete the task. In another study, Mark found that workers who switched tasks more and had less focused time in a day felt that they had achieved less.[12]

Not surprisingly, Mark is concerned about the impact of all this switching on productivity and creativity.

As long ago as 2008, she told Fast Company:

> "When people are switching contexts every 10 and half
> minutes they can't possibly be thinking deeply. There's no
> way people can achieve flow. When I write a research article,
> it takes me a couple of hours before I can even begin to think
> creatively. If I was switching every 10 and half minutes,
> there's just no way I'd be able to think deeply about what I'm
> doing. This is really bad for innovation. When you're on the
> treadmill like this, it's just not possible to achieve flow."[13]

And it's not only e-mail: employees check their smartphones on average 150 times a day.[14] Achieving anything at work when inundated by e-mail, social media, smartphones, and other technologies requires a Herculean feat of self-discipline.

Researcher Sophie Leroy has found that switching rapidly between tasks results in "attention residue" as the brain continues to work on an old task even after it has flipped to a different one.[15] The rise of communication and collaboration tools leads to so much time processing messages and calls for collaboration that workers are forced to perform their jobs at home at night during quiet hours.[16]

The temptation to rely on electronic communication is understandable: it feels less intrusive and requires less emotional energy, particularly if there is any chance of even mild disagreement in the conversation. But the fragmented nature of modern communication perversely bakes new and painful inefficiencies into our workflows. A long conversation conducted over an office chat application may stretch out over hours, span meetings, and sustain multiple interruptions for other pressing tasks. In calculating actual time spent, it might have been far more efficient to pick up the phone or walk over to a colleague's desk. And the mutual understanding that results from these fractured conversations is less than the understanding arrived at through dedicated, rich conversational flows. (For this reason, Alex has a three-response rule. Once an e-mail or chat has gone back and forth, he prefers to talk in person, on a video call, or in a live voice conversation.)

Ultimately, the switching costs of constantly moving between tasks, the costs to our attention of distractions, the compulsive use of e-mail and other communication or notification systems, the lack of boundaries between work and home, and the inefficiencies and artificiality of electronic communications are now making our jobs much harder to do efficiently.

Worse still, we are often imposing switching costs on our co-workers even when we have no intention of interrupting their attention. Anyone who has sat next to someone at a meeting who is constantly checking e-mail or sending chat or text messages can attest that it's hard not to glance over, even inadvertently. Our brains and eyes are wired to monitor activity. So in meetings when laptop and smartphone usage is unrestricted and open, we end up fighting to maintain our focus even if the laptops and smartphones are not our own and we are trying to restrict our own usage to improve our efficiency and focus. How do these switching costs affect our ability to learn and to gain information?

Researchers at the U.S. Air Force Academy created an experiment to test learning performance using three groups. One group took classes in which computers were prohibited in the classroom. Another group took classes in which computers were allowed (although some students elected not to use them). A third group took classes in which tablets were allowed but had to remain flat on the desk. The result was that students in classrooms from which computers were excluded fared much better and learned more in general than students in classrooms in which tech was allowed. What's more, the performance degradation was not dependent on individuals' tech usage; it applied to all members of the group.[17]

So uncontrolled use of technology may harm innocent bystanders; it is a form of ambient pollution. Some economists now treat this spillover effect as a negative externality, just like pollution. With this in mind, much of the confusion over what appears to be lagging productivity in the face of mass tech adoption begins to make sense. Technology is not an unalloyed net positive and can detract from productivity mightily. A blog post titled "Is the economy suffering a crisis of attention?" written by Dan Nixon, of the Bank of England's staff, argued that technology is retarding productivity and driving us to distraction.[18]

Though saying so may seem controversial, it seems to both of us that people often get less done in an hour now than they used to in past decades. This reduces employee satisfaction, because a sense of fulfillment is one of the top motivators for people at work, and a sense of accomplishment doesn't readily flow from spending your entire day distracted and in meetings. Even when we do get as much done in an hour as before, the ubiquitous tech-driven interruptions add significant stress to the work.

Shut Them All Down: A Tale from the Office

A good friend of ours, Mark, is the chief operating officer (COO) of a fast-growing venture-backed enterprise technology start-up. He is a very hands-on manager and a really good guy. We told Mark what we were writing about, and he told us about his notification-detox routine for new staff members:

> Whenever we onboard a new employee, I have to tell them turn off notifications. They come in and want to be notified of every Slack message, be included on every e-mail, and set up those systems to notify them on text if anything comes up around the topics they are interested in. It's borderline insanity.
>
> So I literally sit with them and have them turn off their notifications one by one. We tell them that if we really need to

talk to them on the weekend, we will call them. Most of them seem shocked because it's the first time in memory they have had an employer ask them to connect less, not more. But we have found that too many notifications distract our engineers from doing good work. They can't see it, but we can.

Mark found that by decreasing the communication responsibilities of his workforce, he allowed them to focus more energy on what's important. A rich body of work supports this finding—and finds that this type of focus makes for a happier workforce. To date most of the research has focused on e-mail. You may recall that Gloria Mark has shown that e-mail distractions and overuse lead to greater frustration and stress when performing office tasks.[19] In that particular study, Gloria Mark recommends that workers find ways to limit their exposure to e-mails and to batch e-mail reading into predictable windows.[20] But that's easier said than done. In the next section, you will see why.

In Practice: How Technology Kills
Productivity and Work Satisfaction

Easy access to a multiplicity of tools of text communication is creating unprecedented bottlenecks in the modern employee's workflow. In the past, a message would be a note on the desk or a voicemail. Then came e-mail, and suddenly the volume of messages soared into the dozens per day. Then came chat applications and instant messaging, and those messages—all demanding attention and even more open-ended and less formal—ballooned further. Adding text messages from cell phones or WhatsApp or Skype or Zoom or Facebook Messenger (which your overseas colleagues probably use) commonly leads an employee to have to process hundreds of messages of various types every day and thousands per month; and managers have to deal with five or ten times that number. It's a recipe for message paralysis, courtesy of modern technology.

Collaboration is another way in which technology does us a disservice. In theory, collaboration is a good thing. But it is like most good things: too much of it hurts an organization. In the past, the cost of collaboration was fairly high—requiring meetings, input in person or in writing (typed or handwritten), and other physical actions. Rarely did you see situations in which ten or twenty individuals were consulted on decisions. The miracle of modern technology has made it easier to collaborate on most processes. Want to bring everyone into a decision? Throw up a poll on Slack and ask everyone to weigh in. Or send a group e-mail or group comment request with a link to a Google doc. It's easy, and everyone's feedback can be captured and collated, instantly.

Although this process can add useful inputs and creative thoughts, it can also impede work and increase bureaucracy. Sixty percent of twenty-three thousand employees that the technology consultancy CEB surveyed had to consult with at least ten colleagues, and 30 percent, with at least twenty colleagues, each day in order to effectively discharge their daily duties.[21] Unsurprisingly, CEB has found in surveys that many types of business processes, from hiring new employees to completing enterprise sales to finishing an information technology (IT) project, take longer than they used to a decade earlier.

This retardation partly results from meeting sprawl. In the pre-Internet era, setting up a meeting with five participants was time-consuming. No one had online calendars. There was no easy way for all five to communicate simultaneously until they were face to face, so if they happened to be in different offices, then just setting up the meeting was complicated. These barriers had the effect of limiting the number of meetings that people arranged. As Michael Mankins explains, the introduction of online calendaring programs and e-mails has made it much easier to sync calendars and set up meetings.[22] As a result, by his calculations, the number of meetings has increased dramatically.

Using people analytics and data-mining tools, Malkins and his Bain & Company colleagues combed through e-mails, calendars, chat applications, and other sources of information on how an organization spends its time. They found that the proportion of an organization's time spent in meetings had increased every year from 2008 to 2016, when it had reached roughly 15 percent; and that most mid-level managers spend even more time in meetings: twenty-one hours per week. With an additional eleven hours spent in processing e-mail, the poor manager has less than fifteen hours per week to do thoughtful, deep work![23]

This was just one study. But in our experience, the study understates the case. We have both seen jobs in which senior managers spent more than 70 percent of their time in meetings and most of the rest in interacting by e-mail. And it's worth remembering that the attendance of senior managers in meetings entails a cascade of organizational effects: others must prepare the executives for them. One regular executive committee meeting that cost the executive team seven thousand person-hours per year cost the entire organization three hundred thousand person-hours per year due to meetings cascading from that top-level conclave, and that tally did not include meeting preparation time or research.[24] Work increasingly means conducting meetings, preparing for meetings, and sending and responding to e-mails, with diminishing time available for performing primary job functions. Harris Interactive surveyed two thousand office workers and found that only 45 percent of their time was spent completing their primary job duties; 55 percent was spent on secondary organizational tasks such as e-mail, meetings, and scheduling.[25]

The Slot Machine at the Office

One of the signature trends of technology in the Internet age has been the reversal of technology adoption flows. In the past, the copy machine, the fax, the mobile phone (before smartphones), and the personal computer all started as work tools and then moved into the consumer realm. With the Internet, and with smartphones, that trend reversed. Unexpectedly, consumer tools such as chat, e-mail, and social networks were brought into the workplace—not by IT managers, but by employees looking to increase their productivity. This path had been greased by the demands of workers that they be able to use their own smartphones (and, to a lesser degree, laptops and tablets) to conduct work business such as making phone calls and sending e-mails.

So the slot machine in our pockets was tossed into the workplace, with unsurprising results. Our work tools began to more closely resemble our consumer products. Chat tool Slack uses numerous techniques (as discussed in chapter 1) that encourage workers to pay attention to it as much as possible and consume as much as possible. The company's tagline, after all, is "Where Work Happens." Translation? Don't leave Slack; you will miss something and fail at your job. Urging us to turn on desktop notifications, e-mail notifications when someone mentions our name, and shortcuts that allow us to post GIFs in the chat channel, the product designers of Slack have clearly read Nir Eyal's book.

Slack is one of many services engaging businesses and work teams using approaches similar to those of consumer product designers. In fact, most providers of work technologies, from human resources systems to document-sharing systems to systems managing customer relationships, now emphasize some sort of interruptive notifications system to alert us to a new message or some other event. The result is a blizzard of notifications and intense pressure to keep many of those notification systems on because ignoring a notification is likely to mean ignoring something that somebody considers important.

This new reality of notification insanity obstructs our

concentration not only as individuals but also when we are together—in the flesh, or in a virtual conference. In a study of 1,200 office employees in 2015, videoconferencing company Highfive found that, on average, 4.73 messages, texts, or e-mails are sent by each person during a normal in-person meeting. Seventy-three percent of millennial respondents acknowledged checking their phones during conference calls, and 45 percent acknowledged checking them during in-person meetings. Ironically, 47 percent of respondents' biggest problem with meetings was that co-workers were not paying attention.[26]

So the behavior that intermittent variable rewards induces bleeds over easily from our personal lives into our work lives. The slot machine is in our pockets, on our laptops, and on our videoconferencing systems.

The Trouble with Electronic Communications: Why Can't We Just Talk?

As a result of this trend toward meeting sprawl and a glut of messages, we spend less time in the most important form of communication: one-to-one or very-small-group discussions face to face. The evidence is clear that e-mail is a less effective communication tool than voice or in-person conversations. According to a large body of research, we as humans tend to misunderstand the effect of our e-mails on others and struggle to properly ascertain the tone of those we receive, a typical difficulty being that something intended to be funny may come across as sarcastic.

Research indicates that e-mail is not only more confusing but also less effective than conversation. A study of more than four hundred people given the task of convincing others to take a survey found that in-person requests to strangers were up to thirty-four times as likely to work as e-mailed ones.[27] This experiment clearly is not 100 percent transferrable to the workplace, where workers are all dealing with bosses and co-workers whom they already know, but the effect probably exists there, too. Many of us have had the experience of walking around from desk to desk to collect money for a gift for someone's birthday or sitting in a room together to complete an HR survey after everyone failed to complete it on their own.

Taking this a step further, many technology companies are rethinking whether remote work is as good as in-person work. In March 2017, Michelle Peluso, chief marketing officer of IBM, notified her team of 2,600 distributed employees that they would need to either commute to one of six core office locations or relocate.[28] IBM had long been a pioneer of remote work, so this announcement sent shockwaves through both the company and the technology and business ecosystem.

The announcement, though, was bowing to reality. Most of the research on virtual teams finds them to be less effective when face-to-face collaboration is a core part of the job. As a 2016 article from the *Annual Review of Organizational Psychology and Organizational Behavior* concludes, "Virtual teams have higher levels of confusion and lower levels of satisfaction than their face-to-face counterparts, as well as less accuracy recording their decisions. The end result of these communication problems is a reduction in mutual knowledge among team members. A further issue in virtual teams is a lack of social and status cues."[29]

IBM's Peluso is not alone among big-tech-company executives in having come to realize that remote work is a poor substitute for physical presence. Facebook chairman and CEO Mark Zuckerberg offers employees hefty subsidies to live within a short drive of a Facebook office. Former Apple boss Steve Jobs always emphasized the importance that coming in to the office held. And Google has long paid for an expensive employees' bus service to make it easier for its employees to travel to work.

Academic research is also finding that face-to-face work can result in higher levels of creativity. "If you are part of a team or managing others, you need to be in work [in the office] most days of the week," says economist Nicholas Bloom in an article in the online publication *Quartz*. He should know. Bloom authored one of the most influential papers that originally found strong benefits in telecommuting. The verdict is now clear: the majority of researchers find that remote workforces are not as productive or creative as in-person teams. After decades of favoring remote work, many companies are now mandating at least some physical presence in an office.[30] Technology has failed to conquer distance.

Economists are also struggling to find evidence that technology adoption has led to higher productivity over the past decade. In seminal research on growth and the economy, economist Robert Solow showed that increases in human living standards come largely from improvements in productivity rather than from working longer hours. For many years, productivity improvements driven by technology boosted living standards. But U.S. productivity growth slowed sharply after the Internet boom. During those boom years, from 1995 to 2004, labor productivity rose at an average rate of 3.25 percent per annum, according to Bureau of Labor Statistics data. Between 2004 and 2010 the rate of annual productivity growth leveled to 2 percent on average. Since then, from 2010 through 2016, productivity growth in the U.S. economy has plunged to 0.5 percent, a rate of growth so low that it could mean real declines in living standards. All the other developed economies have experienced a similar deceleration in productivity growth.[31]

This so-called productivity paradox perplexes economists, given the rapid improvements in technology accompanied by plunging prices of tech goods and services. In the past, such innovation would have spurred productivity growth. One common explanation is that we are not measuring productivity gains effectively because our measurement metrics are outdated and still geared to a world of agricultural and industrial production. A second explanation is that, like other massive technological shifts, there is an interim period before businesses and organizations can absorb the changes and take advantage of them to boost productivity (and subsequently profits and wages).

Economist Robert Gordon offers an alternative explanation for technology's failure to continue making marked improvements in productivity. He hypothesizes that earlier technological revolutions—the power grid, the railroad, the automobile—were more foundational and that our digital revolution is less general purpose and more cosmetic.[32] Either the common explanations—poor measurement, slow absorption—or Gordon's interpretation, however, may be consistent with what we propose: that technology has become more destructive and less helpful to work and productivity. The more tech we get, the less productive we are.

Even as technology has had profoundly positive effects on collective productivity—think Amazon, Uber, and Airbnb—the costs of our embrace of it to those using these services have been high. We posit that the advantages in using these technology-advantaged platforms may not outweigh the drag of addictive and interruptive technology systems on our individual levels of productivity and happiness. To say that technology has had no positive effect would be inaccurate, but we must consider the distinction between collective cost–benefit and personal cost–benefit comparisons, which too extend into collective effects. Striking the right balance between habit change and addiction, and between efficiency and obsession, is something our society has failed to do.

The Rising Specter of Technostress

Studies have increasingly labeled the negative effect of technology that we are describing here *technostress*. The by-products of this stress are a poor work–life balance, a growing sense of insecurity, and a strong fear of missing out. According to technostress researcher Monideepa Tarafdar, technostress "comes from our feeling forced to multitask rapidly over streams of information from different devices, having to constantly learn how to use ever-changing I.T., and the sense of being tied to our devices with no real divide between work and home."[33]

Tarafdar cites a host of research that underscores the depth of the pathology. Nearly three-quarters of six hundred computer-using professionals surveyed felt that not being constantly connected would put them at a professional disadvantage, and this cohort reported on average twenty-three minutes per day at home responding to e-mails (though given our human tendency to under-report, the real investment is likely to be significantly higher).

The continual disruption inherent in switching from task to task and endlessly learning new technology tools results in less quality time with colleagues, clients, and partner companies. Tarafdar concludes, "The very qualities that make I.T. useful—dependability, convenience, ease of use and quick processing—may also be harming productivity and people's well-being."[34]

Vivek and Alex have seen many friends wake up to the stress and damage that technology can cause to their productivity and their work. The ones who are slaves to technology—working at organizations that demand fast response times and do a poor job of prioritizing communication demands—rapidly burn out and become disillusioned with their jobs.

To conclude this section on work, we propose a slight twist on an old adage. Parkinson's law, first appearing in a humorous essay published in *The Economist* in 1955, expresses mathematically the notion that work expands so as to fill the time available for its completion.[35] Generally, it is used to describe the sprawl of bureaucracy in an organization, and the phenomenon it describes is never considered a positive thing. A corresponding rule of thumb in the technology age may be the Law of Application Sprawl: the number of applications that employees are asked to use at work is inversely correlated with their productivity and happiness.

Complexity Begets Unhappiness at Work

Much of what has been written about technology implies that modern technology empowers workers to be more effective and more efficient in two ways: by allowing them to be more independent and by providing the means and capacities they need through technological intermediaries. In our experience, the opposite is usually true. We used to share information in the office through the interoffice memo or a conference call; then the channels expanded to encompass e-mail, chat, text, and other technologies. We used to store information on paper in a file in a cabinet. Then we had a hard drive and maybe a network drive. Now, many workers rely on Google Drive, Dropbox, and Box.net as well as individual hard drives and a network drive.

When Alex started work at Mozilla, he found that his

team was using more than a dozen tools for chat, file sharing, and project management, including Skype, IRC, Slack, Dropbox, Campfire, Google Drive, Google Sheets, Google Sidles, and Microsoft Excel. Finding a document or understanding where it was located was next to impossible without finding the person who had the "tribal knowledge." The difficulties were compounded by generic titles on files and a lack of version control.

This was a classic example of chaotic application sprawl. Each application required its own password and its own account management. Each had a different user interface. Each had its own conventions for searching, for sharing, and for other key functions. Some of these conventions may have been similar, but rarely were two functions located in the same place or expressed in the same way.

Each additional application requires employees to remember and deal with more such variations. Larger variations cause bigger problems still. For example, Google Drive, Skype, and Dropbox can all be used for file sharing, but they do it in different ways and have different size limitations on file transfers. So if you want to transfer a file, you have to think which one will be the best tool rather than just share the file. That's not to say that any one of them is wrong. But an employee having to deal with all three must remember three times as many combinations of characteristics, and that becomes taxing very quickly. Google Sheets and Microsoft Excel likewise differ in ways that can make sharing files back and forth a chore. Equally difficult are the dramatic differences between the two in where features are located, and in what features are available. Being proficient at both is a waste of mental energy and time.

Adding external agencies and partners into the mix can exacerbate this problem. Sometimes, in order to be more efficient, the relationship owner inside the client company chooses to use the tools the agency likes rather than the company's usual tools—with the perverse effect of forcing company colleagues to work with yet another application.

Information-retrieval problems are magnified by application sprawl. If you work at a company and need to find the latest version of a document or a presentation, even moderate application sprawl may triple the time it takes for you to search in the different places and applications where the document may be stored. If you received the information but didn't bother to save it to one of your own storage applications, you will also have to search your e-mail and potentially your Slack channels. This replication generates massive overhead. Consulting firm McKinsey & Company and others have found that employees spend 19 percent of their workweek searching for information necessary for doing their jobs. (Ironically, the McKinsey report argues that social tools can reduce the time that employees spend in searching—if they are properly used.)[36]

Application sprawl has become far more pronounced in the past decade because we can now deliver applications over the Internet and have them running in minutes, making the immediate cost of solving a problem with a new application (or tool) very low. Most organizations (and knowledge workers) fail to consider not only the long-term cost, in time and complexity, of any single new application but also the time and complexity inherent in the constellation of tools an organization expects workers to use.

Though technology robs us of work satisfaction partly through the blurring of boundaries between work and home, and partly through consumerization of work technology with intermittent rewards, a considerable portion of the blame for our increasing frustrations at work belongs to poor system and process design. As anyone who has had the experience of reformatting an Excel document into sheets or trying to find a file among a dozen versions on deadline at 11 p.m. on a weeknight can tell you, technology system and process design has real consequences for our daily lives. Yet it's usually tackled only as an afterthought.

The shortfalls in design result in chaos. Nature abhors a vacuum, so work expands to fill time, and so does technology, leading us to spend less time on doing meaningful work and more on managing technology and trying to remember where we stashed our work—not a fulfilling way to spend our working hours. We hate to manage e-mails. We hate to search for files. We spend far more of our day than we'd like to as slaves to our applications, administering technology, attending meetings facilitated by technology, and responding to low-priority communications made frictionless by technology. And we now find that this wonderful technology, rather than make our lives easier, has trapped us in a relentless and demoralizing techno-prison of our own design at the office.

Online Technology and Play

We spend more and more of our time attached to glowing screens. This started with television and has accelerated with the Internet, video games, and smartphones. In this section, we will look at how technology has affected how we play and how we spend leisure time—and at how this may be detracting from our health.

Silent Streets, Soaring Screen Time

Alex grew up in a quiet middle-class neighborhood in Baltimore. The streets were lined with tall trees, and the roads were wide and safe for children on bicycles and scooters. Alex spent much of his childhood outside playing with neighborhood friends. Despite the proximity only two miles away of an exceptionally dangerous neighborhood, Alex's parents did not worry about his playing outside on his own or moving about the neighborhood. He was allowed to ride his bike to friends' homes and to the swimming pool, and to walk a mile to his elementary school unsupervised. And most of his friends lived the same way. These were very normal childhoods for the time.

Decades later, Alex lives in a middle-class neighborhood

in California. The streets are lined with trees, and several streams crisscross the woods and run alongside the wide roads and narrow lanes. The hills that overlook the neighborhood are laced with an extensive network of bike lanes and fire roads that make the neighborhood a magnet for recreational cyclists. In fact, prominent magazines have many times named the Bay Area town where Alex lives one of the best places for outdoor living in the country.

The neighborhood abounds with young families. School enrollment has risen more than 30 percent over the past decade. But in the afternoons and on weekends, the streets and playgrounds are mostly empty. Children still play in the local little league and the local soccer club, they go to the swimming pool, and they spend time outside. But they seem to spend a lot less of their time than children used to playing outdoors, and a lot more playing with technology. Study after study supports this observation.

Most studies find that kids are playing outside about 50 percent less than their parents did, despite their parents' understanding of the importance of outdoor unstructured play. In a large study in the United Kingdom, "Almost all (96 percent) of the 1,001 parents with children aged between four and 14 quizzed for the National Trust thought it was important their children had a connection with nature and thought playing outdoors was important for their development. The research found, on average, children were playing outside for just over four hours a week, compared to 8.2 hours a week when the adults questioned were children."[1]

A similar survey of 1,200 parents in the United States conducted by Gallup (for the toy company Melissa & Doug) found that kids played outside for ten hours per week but played with screens 18.6 hours per week. What's more, the parents of children who spent more than three hours per day on screens were 38 percent more likely to worry about their children's academic performance, 67 percent more likely than other parents to worry about their children's stress levels, and 70 percent more likely to worry about their children's ability to get along with others.[2] Other studies have identified considerable parental concern over how much use their children make of screen devices—and vice versa—and in a 2016 study, 56 percent of parents admitted to checking their mobile devices while driving with their children in the car.[3]

In teenagers and tweens (eight- to twelve-year-olds), the high amount of screen time appears to preclude outdoor activities. A 2015 media-use census of more than two thousand kids by Common Sense Media found that teens spend nine hours per day using some form of electronic media (including music) and that tweens spend six hours per day.[4] Lots of teens and tweens still spend less than two hours a day with screens, but the average times are high. It is likely that teens and tweens are playing a lot less pickup basketball or soccer and a lot more Xbox, iPhone, and Hulu.

A large percentage of young children, teens, and tweens have their own smartphones, tablets, or televisions in their bedrooms. Numerous studies have associated the presence of devices or screens in children's bedrooms with an increase in sedentary activity, a reduction in hours of sleep, and a decline in academic performance. More bedroom screen time also tends to mean more exposure to media violence.[5]

Naturally, some former techies have founded a company that charges money to teach children how to play outside again. A start-up called Tinkergarten uses web-based training and curricula to train teachers in how to lead outdoor play.[6]

The evidence is becoming clear. Technology has fundamentally changed how our children play and how they relate to the natural world around them and to each other. This is not a minor shift. For millennia, humans have gone outside to relax, walk, talk, and connect with one another and nature. Research associates outdoor time, in adults and in children, with low stress levels, good heart health, exercise and activity, and generally positive views of the world.[7] Hospital gardens promote healing, for example, and people who walk in forests have lower blood pressure and lower levels of the stress hormone cortisol in their blood than others do.[8,9] We are letting technology cut our connection to nature.

Because these technologies are so new, no one has yet been able to study their long-term effects on our lives. That said, some scientists and researchers hypothesize that increasing levels of anxiety and depression may be related to spending more time with screens and spending less time walking in the woods, riding bikes, and pursuing outdoor activities in general.

By changing our play lives, technology may also be cutting into our sleep. People who spend time outdoors or in direct sunlight tend to fall asleep earlier and enjoy higher-quality sleep.[10] In adults, the rise of sedentary screen-mediated behaviors is probably a larger factor in the rapid rise in obesity than our bad diets are.[11] This is likely to apply to children as well.

No Vacation from Tech

As you'll recall from the introduction, Vivek's lack of e-mail access while on vacation caused him considerable stress. For the CEO of a start-up, that would be normal; CEOs of U.S. start-ups generally don't take disconnected vacations. The rise of the smartphone and ubiquitous broadband have made it easier than ever to connect to the office to keep up with e-mails, chat messages, and other tasks. As more and more people use their personal devices—in particular, smartphones—for work, the boundary between work and time off blurs. According to McKinsey & Company, 80 percent of employees used their own smartphones for work in 2012.[12] The percentage in what McKinsey terms high-performing companies is even higher.[13]

Of course, this loss of boundaries is in many ways a self-inflicted wound. The iPhone ushered in the BYOD (bring your own device) phenomenon, which most of us have actively embraced. BYOD does give us more control over our devices and obviates the painfully inefficient splitting of our time between two phones. But it also exacerbates the pull of intermittent variable rewards while on vacation, when we now need the mental fortitude to shrug off not only personal social media and e-mail but also the siren song of FOMO and all the connections of work that now live on our smartphones.

Our bosses are part of the problem. Most companies profess that they want their employees to disconnect on vacation. Few have a formal disconnection policy or tech back-stops in place to prevent workers from checking e-mails and Slack. Not surprisingly, workers rarely feel they can disconnect completely. According to a survey conducted by Intel of 13,960 workers between the ages of twenty-one and fifty-four and evenly split between men and women, 55 percent of respondents reported that they had intended to unplug on their vacations but had been unable to do so.[14]

Inability to unplug on vacation is harming our ability to rest, relax, and connect. In the Intel survey and in many others, people who had unplugged said that they had enjoyed their vacations more and were better able to absorb their surroundings. The science of vacations and the impact of tech remains lightly researched, but some studies have found that too much contact with screen technology compromises our ability to recall the vacation. In a 2016 joint study of 713 adults from six countries, vacationers who spent more than two hours per day on their smartphones were 26 percent more likely to struggle to remember their vacations than lighter users of the devices. Working for an hour or more a day on vacation translated into a 43 percent greater chance of inability to remember the trip than in those who worked for less time. Respondents who had used laptops struggled the most to recall their trips.[15]

And what of those video zealots who insist on snapping selfies and videos of their entire trips? "You're taking yourself out of the vacation," says Art Markman, a professor at the University of Texas and the lead author of a study on the topic. "If you're using your phone that much, you're not engaged with your surroundings. There is a qualitative difference between seeing something pretty and snapping a picture of it versus walking around staring at a screen."[16] Of course, taking pictures while on vacation actually helps us remember not only the moment but also its context. But those types of interventions, because they are temporary and intermittent, do not detract from the overall experience and memory.

Love of Nature Becomes Love of Electronic Media

In 2006, biologist Oliver Pergams published a report tracking a clear decline, over three decades, in attendance at national parks, and a decline in requests for licenses for fishing and similar activities.[17] The researchers subsequently analyzed the data across sixteen long-term trends in nature participation beginning in 1987 in the United States and Japan. They found that public participation in a host of nature-based activities in the national parks, including fishing, hunting, and camping, had declined by between 18 percent and 25 percent in just sixteen to eighteen years. Pergams says, "All major lines of evidence point to a general and fundamental shift away from people's participation in nature-based recreation."[18]

Pergams concludes that this decline, viewed as a massive longitudinal dataset, most likely signifies a broad societal shift away from spending leisure time outside. He attributes it at least partly to the rise of multiple forms of electronic media.

According to Outdoor Foundation annual surveys of outdoor activities, the percentage of Americans participating in outdoor activities has trended only slightly downward over all, from 50 percent in 2007 to 48.8 percent in 2017.[19] But a host of studies have found that humans are growing weaker and less physically fit, possibly because of the continuing decline in outdoor activity and leisure-time activity. A meta-analysis in 2013 of fifty studies covering twenty-five million children in twenty-eight countries revealed that children then took ninety seconds longer to complete a one-mile run than had children thirty years earlier.[20] The researchers chose the mile measure because it is a good indicator of aerobic and cardiovascular fitness, both of which are closely related to reductions in heart attack and stroke rates. The researchers found that fitness had declined in the younger generations in nearly every country studied.

Not just aerobic fitness but also strength and flexibility are in decline. "The least fit ten-year-old English child from a class of 30 in 1998 would be one of the five fittest children in the same class tested today," writes Gavin Sandercock, an academic who studies fitness and strength in children.[21] His studies have found that general fitness is declining by nearly 1 percent per year, and that the rate of decline is accelerating.[22] In an earlier study, Sandercock found that the number of sit-ups ten-year-olds can do declined by 27.1 percent between 1998 and 2008. Arm strength fell by 26 percent and grip strength by 7 percent in that period. In 1998, one in twenty children could not hold his or her own weight when hanging from wall bars. In 2008, one in ten could not do so, and another one in ten refused to even try.[23] Researchers have found similar trends in Canadian children.

Sandercock and many other researchers believe that this decline in fitness comes in part from children's spending less time outside playing or just moving around. A 2011 YMCA study of children and their parents found that 58 percent of children between the ages of five and ten played outside fewer than four days a week.[24] A significant percentage of the decline was attributable to the greater convenience to parents of giving children playtime with technology and screen time than of giving them opportunities for outside active play. Not surprisingly, research has shown too that busy tech-using families report lower satisfaction with their leisure time.[25]

These are just a few of the many ways in which pervasive technology is affecting us through our play. The negative effects appear to be accelerating in the younger generations, and we cannot really foretell their ultimate effects. It seems that many kids will never climb a tree, run through a forest, or build a dam in a stream with rocks, and that many adults will never be able to take a vacation without answering work e-mails. For all of us in society, the idea of truly having time for play may be obliterated in our lifetimes unless we can aggressively roll back tech use and guard against the continuing incursion of technology into the most protected parts of our lives.

• 6 •

Online Technology and Life

Are the many discouraging indicators, such as increasing depression and suicide and skyrocketing obesity, actually arising from our use of screen technologies? Clearly, these technologies cannot be the only factor, but in this chapter we look at how our omnipresent screens may be impairing our sleep and undermining other basic pillars of health and entailing a cascade of major compromises of our physical and mental states.

As we were writing this book, many of the tech industry's most prominent members, troubled by the addictive and destructive behaviors that they perceive social media, mobile phones, and other technologies to intentionally foster, began offering serious criticism of the industry. They include former senior executives at Facebook, Google, and other prominent companies. Among the loudest and most insistent was Roger McNamee (whom we later asked to write the foreword to this book). Roger has been investing in technology companies, such as Facebook, for three decades, and introduced Sheryl Sandberg, its present chief operating officer, to its founder, Mark Zuckerberg. From his seat at the table, McNamee has one of the longest perspectives on how the industry is affecting us and our world.

In a *Guardian* interview in October 2017, he pointed out the underlying conundrum: "The people who run Facebook and Google are good people, whose well-intentioned strategies have led to horrific unintended consequences. The problem is that there is nothing the companies can do to address the harm unless they abandon their current advertising models."[1]

When those who have profited the most from tech and nurtured it from birth to dominance come to doubt and fear their creations, it's worth paying very close attention. McNamee feels that these are matters of life and death, and that unbridled technology use is one of the serious threats to humankind.

Time Less Well Spent

When Alex first started using the Internet as a young freelance writer, he was intrigued by the possibility of reading anything from any publication anywhere in the world. He had always been a voracious news consumer, reading multiple magazines a week and several newspapers every day. The Internet made it easy for him to find and read interesting news about anything and everything. For research and to satisfy his curiosity, this was helpful: he often found unexpected associations between articles he was reading and topics he covered. Those led to ideas for articles and new thought paths. And he enjoyed the ability to range freely.

A habit developed. Before bedtime, Alex read the news and browsed the web for cool stories for a bit and then read a physical book or magazine. Reading from paper was calming, and after he read, he fell asleep. Over time, the Internet media consumption gained ground on his dead-tree reading, which fell from two books a week to one a week and then to half a book a week. Then the evening's paper-based reading diminished to primarily magazines in the evening—and then disappeared, replaced by what he found on the Internet.

At the same time, Alex noticed that he was feeling tired much of the time. His bedtime was often pushed back as, after finishing an article, he read another, and then another. His sleep patterns became more erratic. He rarely got a full eight hours of sleep, and he began to regard anything more than six hours as a good night's sleep. He was convinced that this was a normal part of dealing with a busy writing career and that he couldn't turn down opportunities.

Alex was living in Hawaii then, and loved to go surfing in the mornings. As he stayed up later and later each night to read articles—ostensibly to stay on top of the news and learn more about the world—he found that his morning surfing sessions were falling away. Sleep and surfing were two of his greatest joys in life. Yet he chose reading on the Internet over those joys, justifying this change as "work." He was always working (i.e., reading) on the Internet.

In those days, little research had been done into how technology consumption might alter sleep patterns. The temptations then were fewer—there was no binge-watching on Netflix, no Facebook, no Twitter—yet the endless run of text was enough to ensnare Alex. This behavior would come to affect Alex's family life and his marriage. He often stayed up too late "working" on the computer (and sometimes still does). That meant missing early mornings with his children or conversations with his wife. The technology fog enveloped him. He often felt sad and alone, and didn't understand why. From the outside, his life was extraordinary. He lived in Hawaii. He wrote on fascinating topics. He traveled the world. (And yes, he was grateful for his good fortune.) But deep down Alex knew that something was not right.

Beyond the sleep loss, Alex thought often, and still does today, about the sacrifice of precious joys for the dubious benefit of technology consumption. His sense of loneliness and disconnection was probably linked to a life glued to the screen.[2] But the long-term results of sleep deprivation are fairly stark and include a higher risk of obesity, mood disorders, diabetes, cardiac problems, and accelerated mortality.[3]

Alex is hardly alone in his concerns. As we detailed above, leading figures in technology have admitted that they zealously police how their children interact with technology, revealing some strong measures they take to minimize their tech addictions. For example, Justin Rosenstein, the Facebook engineer who envisioned and then created the first Like button, since exiling himself from Snap and Reddit has relied on his executive assistant's use of a parental-control feature to keep him from installing more apps on his iPhone.[4]

Numerous powerful executives from Apple and Google send their young children to Waldorf schools, which ban computers and video games until children are in the seventh or eighth grade.[5] Steve Jobs himself severely limited and policed his children's consumption of screen time.[6] (In fact, Alex's children attended a Waldorf school, and several of their peers' parents worked as high-level executives at high-tech firms, including Google.) If the savviest technology leaders in the world are wary of technology, shouldn't the rest of society exercise more caution and proceed more slowly? If Roger McNamee and the creator of the Facebook Like button are worried, shouldn't everyone be worried?

How Technology Erodes the Core Pillars of Happiness: Sleep and Health

In previous chapters we discussed specifically how technology affects how we love and how we work. These are two of the biggest parts of our lives and two of the largest contributors to our emotional well-being (or lack thereof). And the relationships with technology are fairly direct. For example, too much e-mail and texting during meetings makes the meetings less effective and leaves workers frustrated, and swipe-based online-dating apps can lead to the commoditization of people and to lower self-esteem among heavy users.

We are perhaps even more troubled by the second- and third-order effects of today's technology on our well-being at a deeper level. These are of three types. Risk factors for mortality such as depression, obesity, and diabetes are all strongly associated with sleep deprivation.[7] Loneliness and lack of close connections with other people, also risk factors for mortality and indicators of something deeply amiss in our social lives, arise from behavior that shuts others out.[8] And changes in brain structure arising from the ways we interact with our screens are signaling lower cognitive abilities.

It is one thing to be annoyed and unhappy at work, or to feel unsatisfied in our love lives. It is entirely another thing to say, as we are saying, that technology is undermining our psychological and physiological bedrock.

Tech's Assault on Sleep

Ever since humans have been able to sit up and read at night by candlelight or gas lamp, the night has beckoned. A quiet time to read was heavenly and enveloping. Then came electricity, and it became easier to light the entire house, the town, and the entire city. Night blended into day. Then came television, and we learned how easy it was to sit and stare at a screen rather than read or listen to the radio. Eventually, the Internet linked to television (and later to smartphones) and became our go-to evening entertainment and a staple in our bedrooms. From e-mail to social media to Netflix, endless entertainment—a moveable mobile feast—flowed into our lives.

We were strongly drawn to the flickering light of the screen. A 2011 poll by the National Sleep Foundation, "Communications in the Bedroom," found that 95 percent of respondents watched or looked at some sort of screen within an hour of going to sleep.[9] A 2016 study of 2,750 teenagers in the United Kingdom found that 45 percent admitted to checking their smartphones while they should have been sleeping and 42 percent slept with their phones next to their beds.[10] According to research by venture capital company Accel and data technology company Qualtrics, 53 percent of millennials check e-mail if they wake up in the middle of the night.[11]

Scientists have long known that jobs exposing humans to bright light in the middle of the night—such as graveyard-shift jobs—cause higher rates of heart attacks and other sicknesses, as well as lower life expectancies.[12] The reason, they believe, is that light stimulates our circadian rhythms; the internal biological clock that tells us when to sleep and when to wake. Blue light emitted by electronic screens tends to quash the production of melatonin, a key hormone necessary not only for sleep but also for proper function of the organs. Some sleep researchers and doctors now believe that reduced sleep may also cause obesity and other diseases of modernity due to lowering the metabolic rate during resting times.[13]

What are new are the addictive behaviors in screen consumption and the portability of screens. Prior to the smartphone era, we never turned on the television in the middle of the night when we went to the bathroom, and we didn't sleep with our television sets lying next to us in the bed. And though we may have watched television for hours on end, binge-watching is far more common today, encouraged by features designed into Netflix and other video services. Yes, we got up in the middle of night occasionally to watch television, but it was far less convenient and also far less addictive. Late-night TV was pretty bad, even after the cable television channel explosion. Its ability to grab attention pales by comparison with that of smartphones, which carry not only video services but also social networks, e-mails, text messaging, and the web in a small package that sits conveniently on our nightstand.

Evidence is mounting that binge-watching does indeed cause insomnia and shorten our sleep. In a recent study published in the *Journal of Clinical Sleep Medicine*, 423 adults were surveyed on their screen habits, and 81 percent of them identified themselves as binge watchers. The study found that higher binge-watching frequency was associated with poorer sleep quality, greater fatigue, and insomnia.[14] Because respondents did not find that normal television viewing affects sleep similarly, the authors speculate that the hyper-engaged state occurring during binge-watching is what harms sleep.

Even less-intense exposure to bright screens can reduce sleep. A study by Harvard University researchers found that using an iPad to read at night before going to sleep resulted in a 55 percent fall in melatonin production.[15] The iPad cohort took an average of ten minutes longer than others to fall asleep, and when they did fall asleep, they spent less time in critical REM sleep, which scientists believe to be crucial to restoration. In the morning, those who had viewed the iPads before sleeping felt sleepier and took much longer to begin to feel completely awake and alert. And on that evening, the iPad users began to feel tired a full ninety minutes later than normal. Their circadian rhythms had been scrambled by only a few days of presleep exposure to the devices.

Such circadian disruption may be the constant state of most people today. We do seem to be sleeping less and less over time. According to a 2014 Centers for Disease Control study of hundreds of thousands of people, roughly one in three adult respondents gets less than the recommended seven hours of sleep per night, and one in ten gets less than five hours nightly.[16] These studies are notoriously difficult to vet for accuracy, because self-reporting is always fraught. Nonetheless, a recent study took this question a step further by measuring actual smartphone screen time against sleep-behavior surveys. The findings were clear. More time spent on a smartphone screen correlated with poorer sleep quality and shorter sleep duration. This effect was more pronounced when smartphone use was close to bedtime.[17]

As discussed, sleep deprivation has been clearly associated with a host of health problems, such as depression and most of the diseases of modernity (metabolic disorders, heart problems, and obesity). As the evidence linking both less sleep and poor-quality sleep with more screen time continues to build, more scientists are concluding that our devil's bargain with technology is making us sick and reducing our quality of life—and probably making us more depressed.

Technology Makes Us Lonelier and Sadder

Alex met Steven when they were both living in New York City. They briefly lived in the same neighborhood and shared friends in the media business.

In college, Steven had been active in campus politics and intramural sports. He'd hung out at the local coffee shop where students combined studying for classes with heated political debates. His life was almost a stereotype of what we expect people to do as university students, and he enjoyed the intellectual stimulation.

After graduation, Steven moved to New York City and worked as a Wall Street analyst for a number of years before taking an upper-management role at a company that sold packaged healthy food products. His job matched his ambitions and his aspirations. He liked the idea of helping people be healthy. He was proud of his accomplishments and worked hard.

The long hours at work took a toll. Rather than going out with friends, Steven defaulted to staying home and watching Netflix or Hulu. He stayed connected with his friends from school and home through social media and text messages. He often felt empty and wished he could return to college and the coffee shop. He could go to coffee shops in his city and enjoy a higher-quality, single-estate coffee served by a hipster barista, but all the technology couldn't replace the company of friends.

Even when he was with others, Steven often felt that for him attending to a gadget was more important than the present moment. People always checked smartphones in the middle of dinner or at bars during games of pool or darts, and everyone wearing an Apple Watch was in a constant state of interruption. Steven got the urge to run away whenever he saw a Fitbit on someone's wrist, because he knew the person would check it constantly during any conversation. Through it all, Steven wondered whether it was just he who felt that way or whether his friends—who looked so happy and adventurous on Facebook and who sounded smartly snarky with witty repartee on Twitter—did too.

Steven's concerns echo those voiced by many people Vivek and Alex know—people single and married, young and old, from all over the world. They love their technology. They wouldn't live without it. But the always-on lifestyle makes them lonelier and, ironically, less connected to those around them.

An Epidemic of Loneliness

In 2000, when the World Wide Web was still quite young, political scientist Robert Putnam, in his best-selling social critique *Bowling Alone*, wrote that technology was partly at fault for the growing disconnection of most Americans from friends, family, and society.[18] This was prior to the mass adoption of smartphones, Netflix, Amazon Prime, Instacart, Blue Apron, and all the other services by which technology has encouraged us to fulfill our needs, social and commercial, via a screen rather than by venturing out into the world.

In their book *Loneliness: Human Nature and the Need for Social Connection*, John Cacioppo and William Patrick report that loneliness has increased markedly in the past fifty years.[19] Cross-sectional studies show that between 11 percent and 20 percent of Americans felt lonely in the 1970s and 1980s, depending on the study; those percentages increased to 40 percent to 45 percent by 2010, again depending on the study.[20] (Longitudinal studies gave lower percentages, for reasons that are not yet clear.)

Recent surveys in the United Kingdom also found that more than half the population feels lonely.[21] And people are most likely to be lonely in London, where they live in close proximity to many others and where technology adoption and consumption are extremely high. In the United States, a 2010 survey by the American Association of Retired Persons showed that more than 35 percent of Americans over the age of forty-five feel lonely some or all of the time. That survey also found that heavier users of technology were more likely to feel lonely than lighter users.[22]

The problem is so acute that many leading authorities in public health are calling it "the loneliness epidemic." Many of those experts blame the epidemic, as Putnam did, on technology and social media. Virtual connections are not a replacement for the real friendships that come from face-to-face meetings and the real social interactions we have while out in the world and off our devices.

These findings do not discount the use of online communications in ways that reduce loneliness or build connectedness. When Vivek is travelling in Asia or when Alex is on business in Germany or England, FaceTime or Skype sessions with their families help them to stay connected and close to their loved ones. When the online supplants the real world, though—when the SMS or e-mail becomes the default contact method, when we rarely or almost never spend time outside the tech bubble with people we care about or meet interesting people we would like to know— then it becomes cause for worry.

Chronic loneliness is harmful to your health, shortening life just as obesity and smoking do. People who are lonely or socially isolated or who live by themselves are more than 25 percent more likely than others to die within the next few years.[23]

How large a role does technology play in this epidemic? A host of other causes probably have some influence, including an increase in the number of people living alone and of people living far away from parents, relatives, and the communities in which they grew up. But many surveys show that respondents have strongly implicated the adoption of technology as one of the major catalysts of feelings of loneliness and disconnection.

As discussed, John Cacioppo agrees. So does former surgeon general Vivek Murthy, who has become a leading voice advocating steps to mitigate tech-induced loneliness at work and at home. Murthy said on CBS News in 2017, "Technology can help or hurt; it's simply a tool; but for too many people technology has led to substituting online connections for offline in-person connections, and ultimately I think that has been harmful."[24] And MIT professor and author Sherry Turkle argues in her books *Reclaiming Conversation* and *Alone Together* that technology has alienated us from each other by replacing traditional conversation and face-to-face interactions.[25]

Social-science research offers mixed findings, with some research showing that communications technology use can be beneficial.[26] For example, Facebook use for some is associated with a greater sense of connectedness when it is used primarily for conversation and to lend support to others rather than for envious comparisons. In one study of students in Texas, researchers found a positive correlation between intensity of Facebook use and the subjects' life satisfaction and civic engagement.[27] In another study, researchers found that by asking subjects to increase posting volume on Facebook, the experimentally induced increase in status updating activity reduced loneliness.[28]

That said, the preponderance of studies that look at the pure volume of consumption indicate that isolation and loneliness arise from higher social-media usage. Researchers texted eighty-two subjects five times a day to ask them how Facebook influenced two critical components of subjective well-being: how they were feeling at that moment, and how satisfied they were with their lives. This type of study, called "experience sampling," is a reliable method of measuring real-world behavior and psychological sensations. The researchers found that the more a respondent used Facebook over a two-week stretch, the worse he or she felt at the researchers' next contact. The researchers' conclusion: "On the surface, Facebook provides an invaluable resource for fulfilling the basic human need for social connection. Rather than enhancing well-being, however, these findings suggest that Facebook may undermine it."[29]

A much larger study, published in 2016, examined patterns of social-media consumption among 17,878 people between the ages of nineteen and thirty-two, using a twenty-minute online questionnaire. Respondents were split equally by gender, and 58 percent were Caucasian. More than one-third earned at least $75,000 per year. The survey respondents well represented fairly young adults across boundaries of economics, gender, and race.[30]

The questions focused on how isolated and depressed respondents felt and how frequently they used major social-media platforms such as Facebook, Twitter, LinkedIn, Snapchat, Instagram, and Reddit. The quartile of respondents who checked social media the most frequently (more than fifty times per week) were nearly three times as likely to feel socially isolated as were the quartile of respondents who checked social media the least frequently (only nine times per week). In other words, the study found a strong association between frequency of social media use and feelings of isolation, which strongly correlate with feelings of depression and loneliness.[31]

For Vivek and Alex, this study was eye-opening. Most people they know in media or who are very active in their circles check social media at least a dozen times a day, so nine times a week across the various platforms seems incredibly light.

What about the use of smartphones? People love their

smartphones, but inability to disengage from them diminishes many other aspects of life.[32] Researchers at the University of Essex found that the mere presence of a smartphone can harm the intimacy and quality of conversations.[33] In another study of young adults, researchers found that perception of a romantic partner's higher smartphone dependency correlated with lower satisfaction with the relationship.[34] Another study found that being distracted by a cell phone in the presence of a romantic partner can cause conflict and lead to lower satisfaction with the relationship.[35]

In a highly controversial and widely read article in *The Atlantic*, "Have smartphones destroyed a generation?" sociologist Jean Twenge argues that skyrocketing teenage loneliness and depression since 2011 follows from overexposure to screen communication and underexposure to personal interaction and books.[36] Critics claim that Twenge cherry-picked statistics to form her argument and that teens did not show significant increases in depression. Author and technology journalist Alexandra Samuel comments that "if we've let smartphones run roughshod over our lives, it's not just because they offer respite from our annoying kids, but because they offer respite from our annoying selves."[37]

Your Family on Technology: More Time Together, More Technology, Less Connection

We have all seen it on the playground: a child calling out for the attention of a parent e-mailing or texting or posting on a social network. "Hang on a second, honey!" the parent calls back. In fact, at times in their lives, Alex and Vivek have both been that parent to a degree. Now, through the lens of this book and their research, they see parental distraction as a recurrent theme all around them.

Parents are spending significantly more time with their children than they did three decades ago, families are spending more time together, and spouses are also spending more time with each other. So we should be closer than ever, right? Yet we see evidence that loneliness is increasing across all age groups. What can explain this?

Increasingly, time spent together is also spent engaged with technology. In our society, we spend more time with our families, but for the majority of that time we are also spending time with our devices. The exception may be time spent with children: a few studies show that quality time actually does trump quantity time when it comes to parents' spending time with their children.[38]

We know from research on group dynamics that the mere presence of a cell phone has negative effects on closeness, connection, and conversation quality.[39] As writer Alexandra Samuel points out in her criticism of Twenge's article, social-media use among adults skyrocketed from 2007 to 2010.[40] According to the Pew Research Center, the percentage of Americans between the ages of thirty and forty-nine who use social media rose from 6 percent in 2006 to 86 percent in 2016.[41]

Samuel argues that it's not the kids who are the

problem. Rather, distracted parents are doing even more damage through their inability to disconnect from their own tech. Samuel points to previous research, from the 1980s, that shows that distracted parenting causes significant stress for young children. (In that research, parents conducted another simultaneous activity while caring for their children.) More recent research by Brandon McDaniel and Jenny Radesky found that technological interruptions caused by parents using smartphones are strongly associated with eruption of problem behaviors among young children.[42]

The Internet, along with the technology connected with it, monopolizes attention powerfully enough to erode family closeness, diminishing the family experience on measures that are basic and intuitive. According to a 2008 Pew Internet report, busy tech-using families are less likely to share meals.[43] In another study, researchers tracked seventy-three families for two years to examine the effects of Internet use on social involvement and psychological well-being. The families all used the Internet extensively, and greater use corresponded with less conversation with each other, a smaller social circle of friends around the family, and more depression and loneliness.[44] As we have stated before, this is all very new territory. And, certainly, technology can bring parents and children—and families—closer together in other ways. But without conscious monitoring and understanding of the interactions, technology has the potential to disrupt and harm family interactions.

How Technology Changes Our Brains

Because evidence suggests that even into old age our brains are more plastic than we had ever imagined they were, the effects of technology use on the physical structure and functional systems of our brains are worth examining. Already, clear evidence shows that screen technologies change our fundamental brain structure. London black-cab drivers must pass a daunting test of geospatial acuity called "the Knowledge" before they can receive their licenses. Researchers performed functional magnetic resonance imaging (fMRI) on cabbie trainees prior to their intense studies to acquire the Knowledge and again after they passed the Knowledge test successfully. In the second MRI, the cabbies who had passed showed an increase in the size of the posterior hippocampus, a part of the brain associated with spatial memory.

Subsequent experiments have shown worrying effects of using GPS devices or turn-by-turn software in lieu of navigating by reading maps or using landmarks. Researchers at McGill University in Canada used fMRIs to compare the brain activity of older people who used GPS devices and those who used landmarks to find their way.[45] The researchers noted significantly lower brain activity in the hippocampi of GPS users. They speculated that low brain activity could lead to atrophy of the hippocampus, which might lead to cognitive disorders later in life. Alzheimer's disease is associated with impairments to the hippocampus, causing its sufferers difficulty with spatial orientation and remembering where they are. Researchers also found a greater volume of gray matter in those who used spatial navigation, and this group scored higher on standardized cognition tests. This research suggests that using maps and building cognitive maps is better for the brain over the long term than relying predominantly on GPS.

There are probably other ways in which using tech in lieu of our memory compromises our brain function and potentially our ability to do big-picture thinking. For instance, Nicholas Carr makes the argument that Google search activities have reduced our ability to read deeply and replaced it with a proclivity to skim.[46] He cites a study by researchers from University College London that found that visitors to two popular research sites rarely read more than one or two pages and bounce quickly from one hyperlink to the next.[47] Carr made the controversial argument that Google is actually making us stupid, which we don't believe is true.

In fact, there is plenty of counter-evidence suggesting that Google and heavy Internet usage does have beneficial effects such as enhanced brain function.[48] And humans have long sought to use technology to aid memory, from lists to abacuses for calculation. But we have never before experienced technology that has supplanted reliance on our memory in so many capacities. I don't mean only phone numbers, bank account numbers, and historical snippets; alarmingly, we now refer to YouTube to remind ourselves of skills that previously we fluently recalled: how to clean the lens of a camera, how to dice vegetables, and the like.

And wayfinding and following directions to places fall into this category. Research also implies that we are developing a strong dependence on search engines as the default means of finding information.[49]

It appears that our growing reliance on digital information retrieval is self-reinforcing, making us less and less likely to attempt to recall simple facts rather than Google them. In a study of university students, half of the students were instructed to answer a set of somewhat difficult questions on their own; the other half were instructed to use the Internet. The researchers then gave the students a set of easy questions. Both groups were instructed to use the Internet if they wanted to. Those who had initially found answers on the Internet were more likely to rely on it in answering the easy questions—to which they probably knew the answer. The researchers concluded that this may be an early sign of changes in how we use our memory.[50]

This study dovetails with other research that used brain imaging of habitual Internet users to show that they generate significantly more activity in short-term memory than less-frequent Internet users in performing online tasks.[51] In other words, our brains more easily learn to disregard and discard information we find on line than information we find off line in printed matter. This has serious implications, because long-term memories are necessary to critical thinking and other deep thinking capabilities.

Additionally, this growing dependence has implications for the quality of the information we receive, because search engines' algorithms are not designed to optimize for knowledge but instead are highly commercial vehicles with algorithmic intermediaries whose biases and rules are opaque to us. In short, their primary purpose is to sell our attention to people who want to sell us stuff, rather than to provide us with the information we seek, which is secondary.

As well, when we can use search engines to access information, we humans tend to believe that we know more than in fact we do know.[52] This false confidence has some troubling implications, because narrow information without a connection to a broader understanding of facts and context derived from a good store of relevant memories results in shallower understanding. This, in turn, could make us more likely to fall prey to psychological biases, such as the confirmation bias (the tendency to process information by looking for, or interpreting, information consistent with one's existing beliefs) or recency bias. In others, when we allow search engines to thin our pool of stored facts, then our ability to process and validate information from searches as part of a broader understanding may be diminished and impaired. The limiting aspects of social media, search engines, and smartphones readily explain how our thinking and the information we receive can so easily take on the status of truth. As Eli Parisier outlines in his book, the algorithms and digital intermediaries shape what we find, what we find shapes what we think we know, and what we think we know shapes what we think is the truth.[53]

To be fair, other types of technology and other activities also change our brain structure. Learning foreign languages and studying music both result in structural changes. Certainly the advent of reading from print, and later television, caused some physical changes to our brain structure. Smartphones, though, are new in their ubiquity and in substituting for so much mental activity in our lives. Though that opens up new opportunities to help us—for example, we can quickly learn about health problems—it may also enable our technology exposure to swamp or eradicate beneficial activities, fostering unhealthy reliance and a rewiring of our brains to favor the behavior that search engines, social networks, and video games favor.

Ultimately, it's difficult to say whether technology's influence on our brain structures is affecting our well-being. But we do appear to be undergoing long-term rewirings of our brains in ways we don't understand and have not examined.

A Lesser Version of Our Selves

We've talked here about how misuse of technology can make us unhealthy, ruin our love lives, and come between us and our families. The very, very deepest root of all these ills may well be that technology as we use it today makes us worse people: more narcissistic, less empathetic, less friendly, less caring. Facebook, for instance, is supposed to be about connection and sharing. But Facebook and other social media are equally important for self-presentation, another key human trait. Never before have we had such a perfect platform for self-presentation: infinitely scalable, immediate, and free.[54] And how do we choose to use it? To build Potemkin villages that often mask rather than communicate.

Driving, Texting, Dying

Thousands of people die every year because drivers are texting or using a smartphone while driving.

Director and author Werner Herzog created a half-hour documentary in 2013 for AT&T that portrayed the devastation of families whose loved ones have been killed by distracted drivers. This film, along with others in a series on YouTube, has been viewed tens of millions of times and garnered dozens of news articles. Meanwhile, deaths from distracted driving increased sharply from 2010 to 2015.[55] The poignant films, as well as aggressive "Don't text and drive" campaigns on many state highways, have inexplicably failed to reduce the incidence of this dangerous act.

How is it possible that people can continue to make this choice even in the face of overwhelming evidence of the dangers and constant reminders of the consequences? In the largest distracted-driving study ever undertaken—one that relied on actual cell phone measurements rather than surveys—safe-driving app company Zendrive found in 2017 that drivers used their smartphones in some capacity on 88 percent of driving journeys.[56]

Dying for technology—or killing someone else through our technology addiction—is the ultimate intrusion of tech into our lives. All those who text and drive have the executive ability to stop texting, and even to take an action such as placing their phones in their vehicles' glove compartments or on rear seats—and they know that they should not text while driving.

This problem underscores the ultimate power and peril of the technology we invite into our lives. Distracted driving used to mean talking on the mobile phone while driving. Of course, this is a very dangerous activity. But the rise of texting came after we had already adopted mobile phones. The younger generations no longer spend much time talking on the phone. Texting, by SMS or Snapchat or other means, occupies far more of their time. And texting, which requires both fine manual dexterity and close attention to a small screen, is twenty-three times as dangerous as talking on a phone while driving, according to a study by the Virginia Tech Transportation Institute.[57] The Zendrive study showed that a driver distracted by texting for just two seconds—during which interval a car moving at thirty miles per hour travels twenty-nine yards without driver supervision—increases the likelihood of crashing by 2,000 percent.

We couldn't have easily predicted how commonly drivers would risk their lives in this way, just as we couldn't have easily predicted that young children would soon start gaining access to screens and electronic media in their bedrooms for hours each day and well into the night. Most of the destructive behaviors and consequences of our technology choices sneak up on us. Adoption and participation rates suggest that we as a society continue to hold an overwhelmingly positive view of how technology will affect our lives, even in the face of evidence to the contrary as plain and clear as a blood-stained stretch of highway and the twisted wreckage of a car.

The technology companies are complicit in this sad state of affairs. There are obvious ways to stop people from using smartphones while driving. For example, the car companies could simply block mobile-phone texting or voice usage while a car is moving if there is only one occupant in the car, or allow mobile-phone operation from the passenger seats only. But fixing problems when your customers don't want the fixes takes a real commitment. In fact, vehicle manufacturers are actually moving in the opposite direction, seeking to install Wi-Fi in vehicles. True, technology companies have made modest efforts to restrict smartphone use while driving. Apple added a Do Not Disturb While Driving mode to iOS 11, for example, in the autumn of 2017. But technology companies have for the most part not taken this problem seriously enough to bring about a reduction in distracted driving.

◆ 7 ◆

How Can We Make Technology
Healthier for Humans?

The Blind Men and the Elephant

In a well-known parable, a group of blind men encounters an elephant. Each man touches a different part of the elephant and receives very different tactile feedback. Their later descriptions of the elephant to each other disagree, though each individual's description is accurate and captures one portion of the elephant: a tusk, a leg, an ear. Humans often have only partial information and struggle to understand the feelings and observations of others about the same problem or situation, even though those feelings and observations may be absolutely accurate and valid in that person's context.

Though more multifaceted than our perceptions of an elephant, our relationships with technology are similar: Each of us experiences it differently. Each of us relates to technology in a unique, highly personal way. We lose or cede control, stability, and fulfillment in a million different ways. As Leo Tolstoy wrote in the novel *Anna Karenina*, "All happy families are alike; each unhappy family is unhappy in its own way."

In the same vein, the road back from unhappiness, the

path to taking control over technology, and, by extension, the path to regaining freedom of choice takes a multitude of steps that are different for each of us. The steps nonetheless carry some common characteristics that we can all use as a basis for rediscovering and reentering real life.

Beyond Binary: Rethinking and Redesigning Our Relationship with Technology

The refrain we commonly hear is that we need to unplug and disconnect. Conceptually, this recommendation may feel good as a way to take back total control and to put technology back in its place as a subservient, optional tool. But using technology is no longer a matter of choice.

In San Francisco, if you want to drive across the Golden Gate Bridge, you can no longer use cash to pay the toll. A camera can read your license plate and can send you a bill. But the vast majority of residents pay for the crossing using radio transponders mounted on their dashboards, through a system called FasTrak. Interacting with FasTrak in order to set up and manage your accounts is an online experience. If you choose not to set up a FasTrak account, you have to pay on a per-use basis by check or credit card over the phone. It's a significant inconvenience, and you pay a higher toll.

FasTrak and automated tolls are the way of the future; we can expect that eventually all such transactions will be handled by connected technology. The infrastructure to manage the inner workings of our administrative lives will be digital.

Even in the personal sphere, our friends share pictures digitally. No longer are printed photographs of the soccer team or birthday party mailed to us. Restaurants that use the OpenTable online reservation system often will not take phone calls for reservations; if you want a table, you must reserve it via the Internet. Service after service, business after business, function after function is moving or has moved into the realm of technology in a way that necessitates our participation and connectivity. Yes, we can opt out of those services and businesses, but if we do, we lose out.

An unhealthy relationship with technology is not usually equivalent to alcoholism or drug addiction. With substance abuse, the solution is nearly always abstention and radical changes to life and environment. Drug addicts, for example, are encouraged to move to a different neighborhood in order to avoid old friends from their using days and in order to remove any perceived triggers from their lives. People who believe they have strong online gaming or pornography addictions may be able to similarly move to a new city and avoid the online places that used to ensnare them, but unfortunately such a strategy won't work for technology. We cannot simply stop using technology if we plan to hold good jobs and navigate the world around us. The requirement for us to interface with technology is not decreasing but increasing.

Delta Airlines has just announced, for example, that it will be eliminating its human-attended check-in counters. This means that the only way to get a ticket for a Delta flight will be through a screen of one kind or another. We cannot tell our children that they are not allowed to use mobile phones and tablets if that is how schools administer tests. And we would be hard pressed to stop texting if it is the only or primary means by which our friends and parents contact us.

In the business world, we cannot avoid technology unless we start our own business and run it in some remote town or village. And even that is not realistic. If you were to apply for a white-collar job of any kind and inform the hiring manager that you refuse to use e-mail, you'd get a swift rejection. Likewise, if you refuse to use a smartphone or a mobile phone, you will rule out a wide range of executive roles and other positions for which emergency availability after hours is crucial. And refusing to use videoconferencing on the basis that it makes it too easy to schedule stupid meetings may get you fired.

Real-life practicalities also suggest the value of applying discernment to technology use. The same technology that we may consider unproductive and harmful in one situation may become necessary in another. The cell phone that teenagers cannot put down is a lifeline when they need a ride home from a party because their driver has consumed too much alcohol or taken drugs or simply left without them.

Even the most basic services, such as health care and checking in for a flight, are in line for mandatory digitalization. When Alex needed an X-ray recently, the hospital told him that he had to check in through an online kiosk, an increasingly common requirement. This digitization ranges from the mundane to the mildly entertaining. We have spoons that connect to our smartphones to tell us whether our portion sizes are appropriate. There is the Quirky Egg Minder, a connected egg tray that notifies an app on our phone how many eggs are in our tray. And Brita has made a water-filter pitcher that warns you on line when the water in the pitcher is getting low.

A key part of this digitalization is a move to digitize money. China provides a glimpse into our digital currency and technology future. Use of cash has become a rarity in major cities there.[1] Many merchants simply refuse to accept it, as do many merchants in Sweden, where the government and major banks have collaborated closely to reduce the use of paper money.[2] India, as well, aspires to be cash-free, albeit in an attempt to expose corruption and to ensure equal distribution of government benefits to all its eligible citizens.[3] To make a cashless society possible, India has rolled out a national biometric system that many critics fear has already undermined citizen privacy. Called Aardhar, the system has already been hacked several times but is firmly entrenched in India.[4] Aardhar forces more online banking and the use of smartphones to pay for goods and services or simply lend and repay money.

In Estonia a massive cyberattack by Russia froze the country's infrastructure and convinced the government of this small Baltic country to set up an entirely virtual infrastructure that could be moved to data centers in other countries in the event of a reprise of the attack. This move to cloud computing included critical government and administrative functions, such as the issuance of passports.[5] Other governments are studying what Estonia did and are considering adopting similar models in order to secure business continuity.

So, increasingly, unplugging wholesale is not an option. Nor for most of us is it an appropriate response to life in the age of technology. The question then becomes how to selectively unplug. How can we set better limits? How can we control our environments at work and at home, and the environments our children live in, in order to make them a bulwark against assaults on our freedoms, privacy, and sociability?

"One Small Change I Made That
Improved My Daily Mental State"

Mark Suster is a prominent venture capitalist who has built a huge following for his blog, *Both Sides of the Table*. He is a featured speaker at conferences and a sought-after investor because of the online persona and presence he has built. Suster is a leading voice on critical moral matters facing the venture-capital community and the technology community at large.

In October 2017, Suster published a blog that had a click-bait headline suited to *Buzzfeed* or *HuffPost*: "One small change I made that improved my daily mental state."[6] In the blog, he told of several small changes that he had made to restructure his life in ways that gave him greater control and awareness of his technology consumption.

Two years ago, Suster stopped bringing his smartphone into the bedroom in order to check texts and social feeds before he slept and to check e-mail in the morning. He preferred to have quiet time to think and zone out. Suster writes that in the morning "if I need to get work done I'm infinitely more productive if I come to my computer with a big screen and a keyboard. . . . So my goal was to either have more time to just think or *relax* or admit that I have work to do and do it more productively. I'm happy to say that this has been a huge improvement in my life and productivity."

In the summer of 2017, Suster took a bigger step: he deleted both Facebook and Twitter from his smartphone. He did not stop using the services entirely; he just removed them from his phone to discourage impulsive checking of them. For someone who was prominent in the Twittersphere, often involved in broad-ranging online discussions covering hot topics on technology, this step may seem radical.

Truth be told, Suster began to question the value and wisdom of his entire pattern of technology use. He noticed that he used Facebook and Twitter on mobile primarily when he had little else to do—say, when he found a speaker at a school event boring or when he was waiting for a meeting—and not for anything terribly useful or memorable. He found that it made him less present in the moment. Like many of us, Suster felt bad about the compulsion to use social media even when, deep down, he didn't really want to.

After deleting his phone's social-media applications, Suster found that his days became a lot more enjoyable and productive. He still checks Facebook and Twitter, but he does so a lot less often. He stopped using Facebook for news and started going directly to news sources that he likes and trusts, such as Axios and the *New York Times*. He still uses Facebook, but only for connecting with friends and family and seeing their pictures. He uses Twitter for light professional conversations and to check feeds of companies he is interested in. In general, he uses mobile social applications significantly less. He also shuts off all applications' notifications on his phone. He writes, "Your app can't try to pop my dopamine and try to drag me into being addicted to using it. I'll check when I'm ready." And it's pretty obvious from the post that Mark Suster is happier with his life and his day-to-day existence than he was before making these changes.

Suster took control of his environment and set boundaries and rules. He examined what he liked and what he disliked about how technology was affecting him and thought about how to improve his interactions with it. He put in place a system to design his interactions with technology to suit his needs and maximize the value of his time.

Suster's life modifications may not work, or be necessary, for everyone. Those who check Facebook only twice a day may not feel compelled to erase the application from their phones. People who need to check Facebook every fifteen minutes for their jobs but never check it at night may likewise have no cause for concern about having the application on theirs. But for those who, as Suster did, find themselves using social-media applications from compulsion and feeling unhappy about it, erasing them and changing their environment is a smart strategy.

Designing one's environment to maximize specific

behaviors and outcomes is a well-known strategy in addiction therapy and in the new and burgeoning field of performance management. All of it, in fact, harks back to behavioral design and Fogg's calculus for inducing new behaviors. In his best-selling book *The Power of Habit*, author Charles Duhigg discusses behavioral design in the form of breaking and forming habits.[7] In their book *Nudge*, Cass Sunstein and Richard Thaler, the winners of a Nobel Prize in economics, discuss how to encourage healthy behaviors such as saving money for retirement by employing the same tricks that phone-application developers apply to make us "Like" and share more.[8] On their popular podcast *Freakonomics*, Steven Dubner and economist Steven Leavitt regularly look at matters relating to behavioral design and performance management.

Our relationship with technology is more complicated than going about forming or breaking habits, though. Yes, we can erase Facebook from our phones. But eliminating e-mail access from them may be impossible, even for someone like Suster who has an extraordinary degree of control over his job and his technology use.

Moving Slow, Moving Fast in China

Vivek first visited China more than a decade ago, before the era of wireless data connections and ubiquitous broadband. He found that he could not book ordinary hotels in advance and that catching a taxi was a nightmare because no one spoke English. He needed to have the concierge write his destination on a piece of paper to hand to the taxi driver, praying that he didn't end up in the wrong part of the city.

When he visited again in 2016, Vivek found that the technology landscape had changed. Everyone had a smartphone with fast information transfer. Booking hotels was easy, as were finding online restaurant reviews and catching cabs. Communication was easier, not because more people spoke English but because real-time translation applications had become so good that the Chinese people could hold slow but functional conversations with Vivek by uttering a phrase into their phones and playing back the English version. This trip was less fraught with stress and uncertainty, thanks to modern technology.

The smartphone became a way to help Vivek make the most of his journey and spend less time on the drudgery of logistics and discovery. He felt more in control, better able to navigate, and more mentally free to experience and be present on the trip rather than worry about where he would stay or eat. And whereas using Google Maps in our hometown takes us away from the present and reduces us to watching the blue dot and remembering a lot less about the journey, the map and general online knowledge are an enormous help to the traveler who visits the hinterlands of China, where navigation is more challenging.

Understanding Our Tech Dependence and Addiction

In almost every case with regard to our use of technology, the context matters. In almost every case, the type of activity also matters. Excessive viewing of pornography is likely to have a more negative impact than excessive checking of social media. Porn is generally consumed alone, not as a social activity, and therefore lacks the countervailing benefits of connection and sharing that social media may offer. An online shopping addiction is probably another serious problem, because the long-term consequences of spending too much money on line in an uncontrolled fashion could be far more serious and devastating to someone's life than spending too much time texting or on online dating sites. Then again, a problem with texting while driving can be the most serious problem of all—one with a tragic outcome. Spending eight hours per day on social media, to be sure, has its own real problems: it's pretty hard to hold down a job or hold meaningful conversations in real life if you are constantly checking Twitter, Facebook, Instagram, and Snapchat.

The nuances of context offer special challenges in building smart strategies for healthy technology use and in shifting our interactions with technology from toxic to measured and beneficial. There is no defined category for technology addiction, but psychiatrists have been debating whether Internet addiction is a real malady. It was not added to the latest version of the *Diagnostic and Statistical Manual of Mental Disorders,* the diagnostic bible of mental health professionals around the world. (Online gaming is a subsection of the gambling-addiction section in that publication.) But a working definition of Internet addiction serves as a useful lens through which to view most technology pathologies. In an article on the topic, psychiatrist Jerald Block broke down Internet addiction into three clear subtypes: sexual preoccupation, excessive gaming, and excessive or uncontrolled e-mail or text messaging.[9] This article was written in 2008, so Block probably had not taken account of social media, then not yet in broad adoption. Social media, online shopping, and video watching would be additional subcategories today.

Regardless of the category, Block's enumeration of the phenomenon's negative influences is relevant to nearly any form of addiction or technology pathology.

The first is excessive use, sometimes associated with a loss of sense of time or an (occasionally fatal) neglect of basic needs such as food, drink, bodily evacuation, and sleep.

The second is some form of withdrawal, including feelings of anger, irritability, tension, or depression when a device is not available or when there is no (or limited) Internet connectivity.

The third is tolerance of and willingness to make alterations or purchases to accommodate the addiction. The tolerance may be to acquiring better computer equipment or more software, to spending more hours of use, or to spending a great deal of money.

The fourth is the negative psychic repercussions stemming from arguments, lying, lack of achievement, social isolation, and fatigue. According to the research cited earlier, the repercussions include depression, anxiety, and loneliness.

With these negative influences in mind, we can propose a simple set of questions to ask ourselves in deciding how to create a more mindful and conscious engagement with our technology.

We start by asking a very simple question: does our interaction or use of the technology make us happy or unhappy? There are many derivatives of this question: Does it make us tense or relaxed? Does it make us anxious or calm? The answer may be "both," and that is okay, but we should consider whether, on balance, an interaction leaves us with good or bad feelings.

Good Tech or Bad Tech: Engagement by Design

One way to address the overall question of how a technology affects you is to go through the following exercise. It is a classic decision-framing exercise, not magic; but being able to count, visualize, and weigh effects and considerations is immensely helpful in undertaking it.

Here is what you do. Write down a particular activity or technology at the top of a sheet of paper. (It is definitely best to do this exercise on paper.) It can be anything relating to screens and technology. Draw a line down the middle of the paper. On the left-hand side, list all the positive things and benefits that you feel this technology or technology-driven behavior brings you. On the right-hand side, list all the negatives.

Be expansive. Consider not only the immediate feelings and effects it brings about but also secondary and tangentially related effects that you can perceive. All are critical in the calculation. If the negatives outweigh the positives, then you'll know that a change will be beneficial.

Because we are so immersed in technology, we can slip into a state of constant and paralyzing introspection. At a remove from these disabling effects, this decision-making technique enables life-supporting self-analysis and action—true executive function such as Mark Suster described in implementing his own personal controls.

Ask yourself: Should you remove Facebook or Twitter from your phone? Should you install an application such as Slack on it? Should you ban screens from your bedroom? Should you turn off the Internet on Sundays and after 8 p.m.? Should you lock your phone in your car's glove compartment? If you consume porn or online gaming, should you completely ban it from your life in order to restore balance? These are some of the decisions you will want to make.

You will also want to examine the secondary effects. For example, Alex has until recently used the music app Spotify to play tunes during his runs and workouts. On its face, this seems to make sense. Research has shown that music can positively affect motivation to work out. Alex really liked the feature on Spotify that matches his running pace with song beats of the same pace.

Then he started to pay attention to how much time it was taking for him to manage Spotify during workouts and how much time it was taking away from the workout. Though not the majority of it, the time was considerable. For example, in a standard weightlifting and calisthenics workout, Alex was spending about three minutes per session to manage songs. In a thirty-minute session on a busy day, that was 10 percent of his time—for no good reason. It was dead time due to technology.

Listening to music on Spotify is surely a net positive: Providing an endless selection of tunes with infinite playlists, it opens up rich new worlds. The service also makes sharing with friends very easy. It allows Alex to expose his children to Bach, Mozart, John Coltrane, and Celia Cruz, all from one easy screen, the same screen from which they hear music by Nicki Minaj, the Gym Class Heroes, and Kendrick Lamar. But this example shows the importance of consciously designing the style of our engagement even with a technology application whose use is, by and large, positive.

In engaging with technology, both Vivek and Alex actively and consciously select and lean toward the contexts and uses in which they find the technology behavior to be largely beneficial and satisfying. Though simple, it's an approach that any of us can make work, simply by asking ourselves relevant questions—and being honest about the feelings and other effects the technology raises in us.

Identifying the Problems Tech Causes
in Our Lives: Six Simple Questions

The simple framework outlined in the previous section for analyzing our interactions with technology is one that's easy to expand on. We can efficiently analyze our interactions with technology, and evaluate their effects, through six questions. The answers can be as simple as a mental checklist, and they are usually obvious and intuitive. It can even be useful to list positives and negatives explicitly. The questions to ask yourself about a technology or application are as follows:

1. Does it make us happier or sadder?

2. Do we need to use it as part of our lives or work?

3. Does it warp our sense of time and place in unhealthy ways?

4. Does it change our behavior?

5. Is our use of it hurting those around us?

6. If we stopped using it, would we really miss it?

As an exercise in analyzing our interactions, we might consider one of the primary technology platforms that we use each day, Facebook, and another example of technology use: texting while driving.

First, we analyze our interactions with Facebook.

Question 1: For Vivek and Alex, the answer is that Facebook creates unhappiness. The answer really depends on an individual's nature, but for both of us, the answer is a definite negative: no happiness here.

Question 2: For Vivek and Alex, the answer is that we do not *need* Facebook: ceasing to use it would cause minimal disruption of our daily lives and equilibrium.

Question 3: For Vivek and Alex, Facebook does warp our sense of time. Aimless scrolling down our newsfeeds has resulted in our spending a lot more time than we intended to in Facebook.

Question 4: For Vivek and Alex, Facebook does not radically alter our behavior. We may waste a little time on it, but the changes are barely noticeable in the context of our days. We know that others spend hours and hours on Facebook, but that's not us.

Question 5: For Vivek and Alex, the answer is that our use of Facebook is not hurting those around us. The answer to this question is somewhat dependent on that to the previous question, but it is still instructive. It is important to take an expansive view of whether our Facebook use is harming those around us; sharing photos and checking status to the detriment of other, more important pursuits, such as spending time with friends and family or completing key work projects, may mean that, yes, our usage is harming those around us.

Question 6: For some people, stopping Facebook would be a loss, because it is their chief means of connecting with people important in their lives. Some grandparents find Facebook the easiest way to keep track of far-flung families. But for Vivek and Alex, Facebook is probably something we could take or leave.

Based on these responses, restricting or deleting Facebook would be a positive move for Vivek and Alex.

Next, we analyze our interactions with texting while driving.

Question 1: For Vivek and Alex, this is a far easier case to consider. Texting while driving definitely makes us sadder rather than happier.

Question 2: For Vivek and Alex, we know it is terrible and that we don't need to do it. There is never a situation in which texting while driving is demanded or necessary.

Question 3: For Vivek and Alex, absolutely it warps our sense of time. As discussed in chapter 6, a moment of reading a sentence or two, two seconds of travel in which we imagine nothing occurs, raises the probability of a crash by 2,000 percent. We think we can react much faster than we actually can, and we think that we are less likely to have accidents while we are texting then we actually are.

Question 4: For Vivek and Alex, indeed it can radically alter our behavior. In how many other cases would we say that we embrace a behavior change that increases our chances of dying by more than 50 percent while on the road?

Question 5: For Vivek and Alex, driving while texting definitely does hurt those around us. Hundreds of thousands of car accidents and pedestrian and cyclist injuries are attributable to this pathological behavior.

Question 6: For Vivek and Alex, we might miss texting while driving for a brief instant if we stopped doing it, but then we would probably be exceptionally relieved.

So, based on these responses, our use of the technology clearly detracts massively from the well-being of Vivek and Alex.

Note that texting while driving is a subset of texting, and that texting in other situations is a different matter for consideration. For us, for instance, judicious, appropriate texting is an efficient way to connect to our spouses and friends and children. So context is important.

We can consider each technology either on its own or in a context. Sometimes the technology is not good for us, and sometimes it's merely the context that's not good for us. For example, Alex made a conscious decision to not post his children's pictures on Facebook, as a way to preserve their control over their identities, and decided not to consume friends' photo uploads because it made him sometimes envious of their fun and adventures. Comparisons, he decided, were not joyful. The sharing was nice, but it was too much for him. Competitive by nature, he saw no reason to inflame needless jealousy of what are, in fact, only the very best portions of his friend's lives.

Oddly, when the sharing bares our lives, warts and all, then the impact may be the reverse—it may stoke empathy and push us toward closeness. The posts that Alex remembers to this day are those of a distant friend sharing her struggle with cancer and, simultaneously, with the loss of a child. A *New York Times* tech columnist, Farhad Manjoo, wrote about his experience with a feature on Instagram and Snapchat called "Stories," which attracts a less polished view into the lives of friends and family. Wrote

Manjoo, "Though Stories is only a few weeks old, I have already learned a lot about my friends. It turns out they do not live in perfect houses—some of theirs are as messy as mine—and don't always have perfectly combed hair. They don't always get things done; they sometimes eat less than stellar-looking food; their kids sometimes misbehave just as much as mine."[10]

A False Dichotomy: To Tech or Not to Tech

A question grew in the mind of economics professor David Laibson: are laptops negative externalities in my classes? In his classes at Harvard University, many students used laptops and tablets, their fingers clattering across the keyboards as they took notes—or perhaps perused social media or read e-mails. A negative externality, as discussed earlier, is a secondary negative effect of an action. For example, industrial manufacturing of plastics may produce a negative externality of pollution. Economists have struggled to place a price on negative externalities, as their effects can be manifold and pervasive. In our daily lives, we face negative externalities all the time; second-hand cigarette smoke is a clear example. Wood smoke from fireplaces in wealthy areas of Califiornia is one of the worst pollutants and another clear negative externality.

To students at Harvard, many of whom are paying tens of thousands of dollars a year for an education, Laibson concluded that the interruption caused by the use of technology in class constitutes a meaningful negative externality.

"The web offers instant gratification that undermines our very good intentions to get the most out of class, and that's all about present bias," said Laibson on the *Freakonomics* podcast. "We go into the classroom, and we are convinced, 'I am going to be a good student.' Suddenly, other things become very appealing and very tempting. We're distracted by those other very gratifying opportunities. Suddenly, we've lost forty-five minutes of the fifty-minute lecture."[11]

Laibson thought that banning laptops from his classes would be controversial (though a number of professors ban them from their classes, and many companies have a laptops-closed policy for meetings). After all, many students do use laptops effectively to take notes more quickly than they could with pen and paper, or to search for supporting information and links on line while listening to a lecture. And even though most evidence to date suggests that students learn less effectively in the presence of devices, Laibson did not want be paternalistic and dictatorial.

So Laibson proposed an innovative solution. He would create two class sections. One section would be technology-free; the other would permit the use of laptops and tablets. The reaction from his class has been quite positive. Laibson surveys students at the end of the year asking, "Did this policy of having these two sections facilitate your learning?" On a 0 to 10 scale, the average rating has been just over 8. "It's about letting people choose for themselves, but letting them choose in a deliberative, thoughtful, careful way at the start of the semester," Laibson told *Freakonomics*.

Laibson hopes to make this same choice available to other classes at Harvard in the not-so-distant future.

As Laibson demonstrates, the choice of using technology need not be an all-or-nothing or one-size-fits-nobody solution. "Disconnection or overload" is a false dichotomy, we believe, in the majority of instances. We can design environments that restore choice and allow consideration and fulfillment of individual preferences.

Vivek teaches students at Carnegie Mellon University's (CMU's) College of Engineering at Silicon Valley about advancing technologies. His classes focused on how those technologies can make it possible to solve the grand challenges of humanity, such as providing healthy food, free energy, health, clean water, and a quality education to everyone on earth. Last semester, his classes were live-streamed to the CMU's Pittsburgh campus and to twenty-nine universities all across Mexico and Peru. These foreign schools were generally lesser-known universities in which the majority of students were the first in their families to attend a university. It was an ambitious experiment to see whether Vivek could simultaneously teach thousands of students. But its success was limited.

The Silicon Valley and Pittsburgh students ranked the class as one of the best, but the Mexicans and Peruvians couldn't keep pace—largely because of the language and cultural barriers. Technology simply couldn't bridge these barriers; a human element was clearly needed. On top of that, a constant chatter during class of messages among students in the class Slack channel became a burden for all students. After students complained to him, Vivek decided to ask all students to shut down the Slack channel during class. The students were told that they could take notes on their laptops but that chats and social media would have to wait until after class.

The lesson Vivek learned was that too much technology can be detrimental—and that technology cannot transcend social and cultural barriers. He is no longer offering the class to students outside Carnegie Mellon University, and he has decided that, beginning in the next semester, he will require students to turn off their laptops and smartphones for the first two hours of the four-hour class. The first two hours consist of lectures by Vivek and his son Tarun, who teaches with him, and by visiting professors. The next two hours consist of discussion, debate, and brainstorming; there is little time to post Facebook messages or review Instagram photos during those last two hours. This is how Vivek designed the environment to create for his students a healthy relationship with technology.

At Mozilla, Alex put in place a "no-meeting Friday" policy for his team, to give them a safe time to turn off Slack, avoid e-mail, and focus on the deep work that would benefit from uninterrupted time and make his team's work more rewarding. The concept of no-meeting days is hardly novel, but it is a surprisingly rare tactic considering the simplicity of this choice.

On a vacation to Hawaii, Alex's friend Helen instituted a simple strategy to save her sanity and preserve the sanctity of her vacation. Her teenage daughter was a heavy Instagrammer, whose every trip turned into an Instagram photo shoot. Helen's husband felt compelled to record everything on video in order to share it immediately with his parents, who replied by text. For her part, Helen felt compelled to check Twitter to keep up with the political back-and-forth in Washington, DC, that was part of her job (and classic FOMO). On their previous vacation, they had logged more screen time than real time. On the first day in Maui, Helen rolled out her vacation policy: "Everyone has to pop their SIM card and leave it in the hotel room," she said. "No exceptions."

After some initial protests, the family agreed to give it a try. They drove the storied road to Hana, stopping at waterfalls, and lounged on a black-sand beach before returning to their hotel, exhausted. The day had been a good one. The family had spent time with technology, for sure. The daughter had taken pictures for later Instagram posts, and the husband had shot videos. But, unconnected to the wider world, the family members took the opportunity to interact with each other. The screens were there in the day but did not dominate. The memories were captured, even enhanced, by technology. For that day, they lived in the present and just with each other. It was a fairy-tale ending, achieved simply by ejecting their SIM cards.

Truly, we can have both our technology and our freedoms. Whether to tech or not to tech is a question easily transcended through creative environmental and behavioral design.

Defensive Tools, Defensive Strategies

The arsenal of hardware and software to help people better control their technology use and make it more mindful is expanding. There are a handful of smartphone applications, such as Freedom, Focus, Unglue, and Moment, that measure active time spent on the phone and limit its use to a specified maximum. Other applications, such as Space by Dopamine Labs, insert a pause, delaying users' access to applications or services they wish to limit their use of. The pause helps people break their habit by making their choice to check Facebook or LinkedIn more conscious and more active. At work, there are any number of applications designed to better structure and batch e-mail usage, such as Inbox When Ready.

There are routers such as Torch and Circle that let parents shut off certain users' Internet access during certain hours of the day and filter the kinds of activity they can engage in. This capability has been around for a while, going back to OpenDNS, which Alex used, but for people who are less comfortable managing complicated networking technologies, the newer products make the process easier for regaining control of their technology. There are, too, parental controls baked into a growing number of the services, including Facebook. Curiously, Instagram defiantly refuses to allow for parental controls, although it is possible to download apps such as Netsanity, which can be used to control a minor's Instagram usage and block the use of other applications.

Software solutions may not be the most effective or the most rewarding ways to control technology use. Layering more technology on top of technology doesn't attack the root of the problem and doesn't help us build the mental muscle that will, over time, help us recover our freedom of choice and our real lives. Using additional technology to corral our own use of technology adds more to the cognitive load we carry by adding a new technology product to our portfolio of things we have to mentally manage. A technology product may work for some people, though, and we encourage you to continue using whatever works.

We prefer solutions to problems of toxic technology use that, like the six questions discussed in this chapter, are more generic and human based. With that in mind, chapter 8 presents a few playbooks that we hope you can use. If they sound simple, that's because they are. We consider that injecting another prophylactic layer of technology management—products to manage products, applications to manage applications—on top of the existing mental burden of managing our technology lives is a waste of our finite mental capacities.

• 8 •

A Vision for a More Humane Tech

Imagine that your smartphone had a pause button that would stop all buzzing and notifications for multiples of fifteen minutes and clear your home screen to leave nothing on it except for a clock—and that you could block all incoming messages by pushing a single button on the phone. You might say that your phone already does that with its Do Not Disturb (DND) mode, but DND requires quite a bit of management in its present state, and when the DND period ends, you get a rush of notifications, followed by newly arriving notifications. What if you could program the phone to send you notifications only on the hour, in regular batches?

In fact, someone has already invented a phone like that. It's named "Siempo," and it was designed by a team from the ground up to encourage more conscious, thoughtful use of applications and technology, and to return to users control over their lives. Siempo calls the device the "phone for humans."[1] Siempo was launched on Kickstarter in March 2017, and it raised only a fraction of its goal of $500,000. Sadly, the market did not fully validate what Siempo was offering.

What Siempo was trying to build is something long argued for by Tristan Harris. Harris says that a standard feature of all technology should be our ability to choose to focus and regain our choice and control over it. That might seem far-fetched, but is it?

Actually, the number of tech companies adopting Harris's philosophy—or at least ending up conforming to his ideals—is growing. These companies are giving us human choices in normal product categories, or they are giving us products that help us restore the humanity to our technology tools. They range from giants such as Apple to one-person start-ups such as Moment. And they range from travel applications to project-management applications to dating applications. Siempo struggled to raise sufficient funding, but many of these freedom-friendly companies are doing very well, with at least one approaching unicorn status (a valuation of $1 billion). These applications replace the focus of consuming and extracting money with one of giving us back our time, giving us intelligent choices that are better aligned with our own interests, and contributing to our well-being.

Apps That Make Tech Better

Some of life's most frustrating moments happen when we're traveling. Flight delays, long layovers, expensive routes, and other miseries compound to make modern travel unpleasant much of the time. Today's travel search engines too are unpleasant. Yes, they pull in loads of information. But the information is poorly presented or confusing, and it's difficult to navigate.

This is why Hipmunk, a popular travel search engine and hotel-booking portal, created an "agony" algorithm. The algorithm calculates an agony ranking for flights based on a number of factors such as flight duration, number of stops, and likelihood of delays and weighs them against price. Travel shoppers on Hipmunk can sort by any other combination of parameters, but many default to results ranked by "agony" by clicking on the agony tab. This is a choice that takes into account our comfort: the human choice.

Alex has long been a lover of Hipmunk and its agony tab, so he was encouraged when he saw that Tristan Harris had given it his thumbs-up. In 2016, Hipmunk was purchased by expense-management company Concur for between $50 and $100 million. Although that wasn't a huge success for investors, Hipmunk demonstrated that human-centric design can succeed in making an online shopping process that we expect to be painfully complex bearable and even fun.

In fact, we are encouraged by the trend of elegant simplicity that we see in a lot of product and service design today. Lyft and Uber are a case in point. Before their advent, a taxicab required a phone call, or a wait on the street for a random cab that we didn't know would come (and that might not come). Uber and Lyft made the entire process transparent: we know when the car is coming, where it's coming from, who the driver is, the license-plate number, and the color, make, and model of the car. We can even text or call the driver. Time previously wasted in hassling with the old system is time we have regained—with the push of a button. Another case in point is the Nest Thermostat, built by a former Apple designer, Tony Fadell: it has made saving energy and managing the temperature of your house simple. And it's easy on the eyes.

To regain control over their browsing experiences, many people we know are using AdBlock or similar services that strip advertising from web pages and minimize invasive, interruptive ads. By some estimates, the proportions of people using ad-blocking services is approaching 40 percent, and with good reason. Another way to handle this is to put a web page into "reader mode," something you can do either in the browser with a feature flag, or in Google Chrome by installing the extension Just Read. Apple's Safari browser has a built-in reader mode. We understand that this puts a lot of pressure on publishers, but they can surely work with it. And we could voluntarily white-list sites and view ads on domains whose content we value and whose publishers we would like to support. By using an ad blocker, we restore the balance of power and put the human user in control.

We should note that the current prevalent advertising-driven economy of the Internet may be at cross-purposes with healthy technology usage. The sole currency of this economy is attention, as measured in impressions, views, clicks, and other metrics of uncontrolled consumption. Unless Facebook, Google, and others can figure out a way to increase their revenues through lower engagement, it seems unlikely that they will themselves create an alternative. Granted, Google has regularly reduced the number of ads showing on a page in order to reduce ad overload, but the goal of this reduction remains unchanged: to obtain more (and more-valuable) clicks, ultimately in support of charging more for the ads.

Publishers have already learned the hard way that the attention economy, in particular when it is controlled by the giant platforms, has no way to support quality of interaction or Tristan Harris's concept of time well spent. That said, a number of quality-centric publishers, such as the *Financial Times*, the *Wall Street Journal*, and the *New York Times*, have built subscription businesses that are overtaking their advertising businesses. The paying users also support rising ad costs through engaging and showing clear desire to be reading that content rather than a drive-by dump of something from a social-media link.

Very encouragingly, a rising chorus has called on Apple to engineer an iPhone that promotes healthier interactions with technology, particularly for children. That may be a challenge, but Apple has made a start. It added "Night Shift" mode to iOS and to its most recent Mac OS in late 2017. Night Shift changes the colors emitted on the screen to reduce the amount of potentially sleep-disturbing blue light that it exposes users to.

Better Work through Tech

The productivity slowdown in the office has not gone unnoticed. Many companies have started using what some call "agile development" to focus technical work (and increasingly marketing work) by reducing the focus of each worker to a single task and unifying teams around a limited set of tools and processes. Many productivity gurus agree that greater focus can be achieved by setting up better processes and creating limits. The most famous of these is David Allen, who developed the GTD (Getting Things Done) method and built it into a mini business empire of seminars, branded products, and curricula. We can't necessarily recommend any of these methods or strategies. Many do have merits, although in our experience, when imposed on teams, they can feel top-heavy and paternalistic.

For individual use, there are dozens of productivity applications designed to help us refocus and get more done by wasting less time on digital frittering. Calendly, Meetingbird, and a number of other applications sync with an online calendar and let you shift the burden of scheduling meetings to the person who is asking for one by letting him or her pick a free time from your calendar without having access to its content. Inbox When Ready lets you set up detailed e-mail rules on when you can access e-mails and how you want the inbox to look, and thus helps you control your urge to use e-mail. SaneBox automatically filters out less-useful mail (promotions, updates, etc.) and raises the most important e-mails to the top of the queue. There are apps and browser extensions to block or ration social-media usage and to track exactly how you spend your time on your computer (breaking down your actions by application to the nearest minute on a daily basis). And applications such as Moment can show which applications you are using on your phone and even poll you on how much you like using them.

These productivity tools fall into three classes: those that seek to improve processes, those that seek to limit distractions, and those that seek to illuminate our habits. All three can be useful, but only if they fit nicely into your working day. Calendaring applications, for example, do not handle situations well when someone is juggling personal and work calendars and mirroring schedules in order to keep a spouse informed. Ad blocking works well until enough publishers start blocking content; then managing the ad blocker becomes a chore in itself. Activity-tracking software works to the extent that it serves to enable conscious changes in behavior, not merely as a vanity metric. All three of these concepts, however, play a critical role in restoring freedom of choice by making our work activities conscious choices, a key contributor to work satisfaction. On the book's website, HackedHappiness.com, we include a page that lists many of these tools.

It is a shame, though, to have to cobble these tools together for want of human-centric product design that would incorporate their capabilities and characteristics. Imagine if Apple and Samsung offered such features as prominent features of their phones. Apple does have some useful features on the iPhone, such as Do Not Disturb and Night Shift, but there is no way to globally set notification preferences for all applications. If Tim Cook told his developers to build DND on steroids for smartphones and laptops, then our freedom of choice and our control over how we use our phones would at least improve.

Notifications and other types of application noise are not the only way in which technology products bludgeon us into confused submission. The advertisements that now dominate many search-engine results—in particular on mobile devices—have rendered information retrieval essentially an act of blind faith answered with results that are purchased by those with the wallet to buy the ads at the top of the page.

What if, as an alternative approach to revenue raising, Google offered its own AdBlock button that let us see search results stripped of all advertising? Even those results would be imperfect: very large sites with large budgets have a lot more time and energy to invest in developing so-called inbound links from other sites—the primary measure of site relevance that Google's search engine algorithms trade in. But at least results would be less readily up for sale. Google might even make up for the initial loss of ad revenue through search results by improving user engagement with the resulting pages. And users who stand to benefit from the ads, as Google claims many do, could continue to search in the normal manner.

Alternatively, as we asked earlier, what if Google offered a paid ad-free version? How much would people pay? Would they pay? How many people would adopt it? We don't know, but it would be a worthwhile experiment. The point is that it's feasible to give users a clean and easy choice and let them decide.

Google did announce such a product when it was faced with competition from Netflix for its video platform, YouTube. The product is a monthly YouTube subscription that removes advertisements from all videos, on every device you and your family watch. It also allows the downloading of videos to phone and tablet and makes them available for up to thirty days to watch without a connection. For $10 a month at the time of this writing, you can hire freedom from the mindless ads that YouTube serves up and can take the videos with you wherever you travel—even if there is no Internet connection. Vivek has been using it and does not miss the incessant ads and interruptions of content.

And some of us would gladly pay Google another $10 a month—or more—to get the same quality of search results as before it began selling ads to the highest bidder. We yearn to see web content unobscured by digital billboards.

Tristan Harris suggests an interesting scenario for Google Maps. What if an application on our device knew or inferred that we want to stay in better shape? Then it might highlight walking options in Google Map directions. And if the application knew that we like podcasts, it might even suggest a podcast from our playlist that perfectly fits the walk's projected duration. In such instances, the application would be working with our own stated preferences to give us choices that reflect our needs and wants and help us live better.

What might Facebook be like if it offered multiple modes? One of these modes would show posts only from your close friends and not consider Facebook's content-selection algorithms. (Of course, you can do this manually, spending hours sifting through friends and indicating which ones you want to follow and which ones you want notification preferences from.) Another mode could select just "trending" news from across Facebook or across your network. An alternative to such modes could be an ad-free paid version of Facebook. Perhaps the lack of a paid version is just another example of the big tech businesses' shortsightedness: Twitter founder Biz Stone advocated for this model in a Medium post four years ago. The paid version could offer additional benefits, such as special commercial offers from stores you chose to receive them from.

Facebook could become the broker for your digital life by making the users the boss and selling access on an opt-in rather than an opt-out basis. This might be less lucrative in the short term, but in the long term, true opt-in businesses based on subscriptions or voluntary participation tend to perform exceptionally well when the product is worth paying for. LinkedIn, for instance, has been highly successful in its lucrative subscription model for professional users, who use it to send so-called InMails and to enhance their ability to search the network.

Beyond Tech Platforms: A Holistic Approach

Beyond Big Technology and the massive platforms that control so much of our online lives lurks another layer that is perhaps more difficult to adapt to a vision of coexistence with a more humane technology. One form of technology noise is the barrage of meeting requests and e-mails we face at work, many of which we accept because we fear the consequences of rejecting them.

Another form of technology noise is the omnipresence of screens, not just in our homes but also in bars, restaurants, and elevators, even at gas-station pumps, where a television advertising network now competes for our attention. In the entertainment media, product placements are another form of invasion, a subtle way to influence our opinions and preferences that, in an age of video games and digital video transmissions, is turning entertainment into yet another advertising venue.

Even *within* our screen time, our attention is becoming increasingly splintered. In the broadcast of a baseball game, there are two and sometimes as many as four split screens showing data analysis, interviews, or other video footage, as well, of course, as advertisements for other programs that we must watch after we finish with this one!

What this comes down to are simple questions: What is our attention worth? How can we force companies to put a value on it as they design their products and their user experiences? Is that even a viable option?

As discussed earlier in this chapter, there are dozens of applications designed to help us avoid overuse and regain some control over our digital lives. Nearly all are only palliatives of one kind or another; they paper over the core problem. With every system upgrade, we generally have to go back and redesign and reinstate our chosen digital lifestyles. That becomes exhausting, so it is no wonder that many of us choose to give up and allow technology to dictate what we see and when we see it or engage with it. There is no easy way to control it. Even if we do like all the features, we would benefit by being able to more easily control them and turn them on or off. Apple solved a huge user-experience problem by introducing the one-button mouse; other ingenious product and application designers can also surely design the user experience to be more humane.

Until they do, we humans are left to remodel our own user experience as best we can, wielding the sledgehammers of brute-force deletions, hard-and-fast rules, and turning devices off for long stretches of the day and night. It's risible to have to take this upon ourselves as technology users, because one would hope that the brilliance of the technology sector would extend to designing a healthy way for tech to coexist with humans, maintaining and even increasing our freedoms, our quality of life, and our control without killing their profits. That said, the next section offers some suggestions toward that end.

How Technology Makers Can and Should
Respond: Restraint, Respect, and Choice

How can we redesign technology to better respect choice, reduce technostress, and foster creative and social fulfillment? The ideal solution would be easy to implement and to customize, and easy to apply to multiple devices and platforms. It would have a centralized user account that allowed you to customize all your interactions and notifications, to which all applications would refer for guidance and permission. It would be, in other words, a true user agent, an intermediary that brokered our attention and implemented our rules in eliciting it. The concept of such a user agent has been discussed repeatedly in industry, but it has never been instituted. Given our growing collective discontent, our epidemic loneliness, and our declining productivity, the time may have come when such a solution is no longer simply ideal but essential.

Ultimately, such an agent will have to be habit-forming technology. It will have to take all the techniques that Silicon Valley's "user-experience designers," say at Facebook and Netflix, have used in forming destructive habits and invert them. We need good magic. We need technology to enhance chronic focus rather than bombard us with chronic distraction; to encourage beneficial habits rather than motivate us to pursue pathological addictions; to promote productivity, connectedness, creativity, spontaneity, and engagement rather than cheap facsimiles of those qualities. The well-lived life, which has never been further from our reach, is one that good technology design could and should make more straightforwardly and universally attainable than ever before.

What Moment, Siempo, Unglue, Calendly, SaneBox, and similar applications are aiming to deliver is that kind of beneficial magic and focus enhancement. They seek to reduce the frictions that we as users must endure in attaining focus. Most of the mechanisms that inhibit or destroy our focus create stress, unhappiness, regret, or sadness once they become too interruptive. (As described earlier, Mark Suster, for example, wrote that he had deleted Twitter from his phone because the constant news stream made him unhappy all too often.)

In sympathy with Tristan Harris's user-rights manifesto, we have a vision of a technology world that works for humans rather than against them and that has each and every company consider the long-term health and benefit of its users to be an imperative design consideration. Even if it meant less profit in the short term, they would restrain themselves from inducing patterns of destructive overconsumption. We propose that this would work as follows.

First, technology makers would define patterns that suggest problem use—preferably without identifying problem users as individuals. Such patterns would include spending an inordinate amount of time with the product, spending too much money, or regularly exhibiting unhealthy behaviors such as binge-watching. Triggered by such patterns in its use, the technology product would treat the user differently, offering help in altering these patterns. This may seem like a patronizing approach, but we would wager that, given the option, many people would welcome the help.

In a work context, we might see Slack warning heavy users not only that they must keep desktop notifications enabled but also that they are in the upper percentage of GIF senders or message senders. E-mail providers might offer batch receiving of messages to its users who otherwise tend to respond the most quickly (which could indicate compulsion to check for and respond to messages). Or every e-mail client could offer batch receiving as its default mode, or simply ask us every day how many times we want to check e-mail that day (in a Siri-like voice, of course).

For consumers of video, product designers for Netflix and YouTube, for example, would make auto-play an opt-in function. In fact, opt in would become the default rather than requiring opt out as the standard product design. And when product designers did choose to deploy opt out, they would allow people to opt out very easily whenever the feature was showing. For example, every video auto-play would also display a Stop Auto-Play button as a preference. That might slow consumption, but then again it might help all of us feel more in control, be more productive, and be more loyal customers.

But how can initiatives such as these be given teeth and profit motive? We are hopeful that, in some cases, the profit motive will take care of itself. Both Netflix and LinkedIn have cracked that nut, as have Spotify and numerous other subscription-based technology businesses. In such a case, inducing massive consumption beyond a certain point becomes counterproductive of customer satisfaction; we suspect that these businesses know exactly where that threshold lies.

And, yes, those platforms are now just as guilty of the same attention-grabbing offenses as the free platforms. But they have the benefit of paid users and a willingness to put a value on attention, participation, and services rendered.

The challenge is to price attention, participation, and customer satisfaction and loyalty in the attention economy.

So how might this work? Imagine that Facebook charges looked like our regular mobile-phone bills with a set of à la carte services. We could opt in or out of those services—for example, no ads in our feed, and a Focus button on our home page that blocks all notifications—and pay for them as features.

We realize that charging users is exceptionally difficult, and is probably not going to happen with Facebook or Google; it will probably be the next entrant that cracks this model. But we can point to one example in corporate America where businesses are showing exceptional ability to put a price on such fuzzy costs: Benefit Corporations (B Corps), whose ranks are growing quickly. Some extremely profitable and successful brands, such as Patagonia and Athleta, have become B Corps. Technology companies would find it at least as simple to enact a similar ethos of ensuring that the product and service on offer does no harm and is in the best interests of society.

The B Corp validation and rating process could easily incorporate a set of values and measurements specifically designed for technology companies. For example, a tech company that could be rated as a B Corp would allow users to unsubscribe from the service in no more than three clicks and without having to send an e-mail or make a phone call. The government of China mandates that game companies put in place user warnings beyond a certain number of hours; B Corp tech companies would have to warn users that their actions were perhaps unhealthy after they averaged more than, say, two hours of use per day over a week.

How Employers Can and Should Respond: Reduce Noise, Tools, and the Rule of Three

In the workplace, the existence of a hierarchy simplifies the rationalization of distractions. Rather than think only of how a new tool or application is going to provide yet another feature and capability, in deciding which tools to use, workers and managers must think about the cumulative effects of their decisions. This may fly in the face of the BYOA (Bring Your Own App) culture, but it acknowledges the collective price of switching costs.

We advocate what we call "the Rule of Three": teams should try to narrow down their primary tools and applications, beyond e-mail, calendar, and word processing, to three choices. This, we believe, will cover the work requirements for 90 percent of teams in the workplace today. A team's need for more than three tools is commonly a sign of distress and trouble (though in certain cases it simply indicates that the tools for the team's job are not integrated). Society is levying high switching costs on that team.

The Silent Start and Other Ways to Rethink Work

Slowing down interruptions and encouraging more deep work is exceptionally difficult when a multitask ethos is ingrained. To combat multitasking during meetings and try to keep meetings meaningful, Amazon mandates that attendees spend the first part of a meeting reading a printed agenda and additional information that sets the table for the meeting. CEO Jeff Bezos calls it "the silent start."[2] This is very good meeting hygiene and, according to Bezos, a wonderful way to spark innovation and interesting ideas.

It is really up to organizational leaders, such as Bezos, to give workers and their organizations the safe space in which to be creative and productive and fulfill their potential. We advocate that every company create a cohesive productivity policy. Some companies are already performing in-depth reviews on employee productivity and practices, meeting structures, meeting attendance, how quickly e-mails are answered, how documents are shared—all the minutiae that make up the bulk of our working day and can easily create employment hell.

The policy—or we can call it a performance enhancement plan (PEP) if you need an HR-friendly buzzword—could signify both management's willingness to allow employees to design their work experiences and the willingness of employees to take ownership for the creation of a healthy environment that promotes productivity, balance, flow, and, as a consequence, work satisfaction.

For example, what if your company had a policy to actively discourage employees from checking and responding to e-mails every five minutes, and instead set up the e-mail servers to batch-receive e-mail messages on the hour or the half-hour? (This is entirely possible and almost trivial to implement.) What if companies had a warning system baked into calendars to alert employees that they are allocating more than 20 percent of their time to meetings, and highlighting transmission and reception of messages on weekends or e-mail threads that extend beyond five round trips? What if the system alerted users who overuse "reply all" in environments that don't generally need it?

We could, in principle, build a work-satisfaction index—a single number that rates an employee's state of work fulfillment and a manager's adherence to these kinds of policies. Using the same tools that increasingly track productivity and monitor what employees do, we could easily make visible the unhealthy practices that lead to so much work frustration and discontentment.

Some companies have ad hoc versions of systems that work by giving employees maximum freedom. Netflix, for example, tells employees to take as much vacation as they want to and as much family leave as they want to, and even to show up at the office when they want to; its only requirement is for employees to be top performers. Best Buy for a while had a "meetings entirely optional policy" as part of its Results Oriented Work Program. That policy was discontinued in 2013, but many employees and managers felt that it worked extremely well by allowing people to design their work engagements and minimize imposition of time-wasting distractions. The core message of all this should be "Design the work environment you need, and we will back you up and let you deliver. Just be sure to deliver."

Building such a culture of freedom takes a particularly strong stomach, for CEOs, managers, and employees. It necessitates that all of us take personal responsibility to banish FOMO, to avoid interruptions, and to silence our inner technology demons. So far, we collectively have remained unsure whether such drastic approaches could work over the long haul, because the broader pressures to be always "on" amid job uncertainty are so great.

As employees, we should always ask what the realistic expectations are for our roles and the culture of the company. If that culture goes against our notion of designed freedom of choice and flow, then we should vote with our feet and find work somewhere else. As leaders, executives must begin paying more than lip service to notions of holistic approaches to employee and organizational health.

For the most part, in a white-collar environment, productivity, engagement, and workplace satisfaction are closely related. Productivity is a measure of the output achieved per hour of input. Note that this is a ratio of one to the other, not an absolute. Numerous studies have shown that beyond a certain point, working excessive hours yields rapidly diminishing returns. True, that balance may shift, for instance when we are under looming project deadlines. But, as when we've crammed for exams at university, we need a break afterward: time to recharge our energy and our well-being. This too must be baked into the work environment.

How Governments Should Respond:
Regulate Where Necessary

In some cases, resolution of the problems resulting from technology design and build may require government intervention. Though we would hope that companies would recognize when what they are doing is clearly not in the public interest, the profit motive may be too deeply ingrained, as has been borne out in other industries that foster deep addictions, such as the tobacco and alcohol industries, the gambling industry, and the processed-food (junk food) industry.

Particularly in order to protect children, we think that regulation or government intervention will be necessary. Some governments have taken that step. South Korea considers Internet addiction one of its most serious public-health issues. The average South Korean high-school student spends twenty-three hours a week playing online games. As of June 2007, South Korea had trained more than one thousand counselors to treat Internet and gaming addiction. The government has also signed up nearly two hundred hospitals and treatment centers in the campaign. In China, current laws strongly discourage more than three hours of daily game use by children under the age of eighteen. Chinese Internet game operators are mandated to install anti-addiction systems whereby the first three hours of play proceed normally, but the games themselves award points far more slowly for the next two hours, after which users receive in-game warnings like this: "You have entered unhealthy game time; please go off line immediately to rest. If you do not, your health will be damaged, and your points will be cut to zero."[3]

These ways of approaching the problem do seem extraordinarily heavy-handed. An example of perhaps a better regulatory approach would be to ban companies from the advertising of games marketed for children under the age of eighteen, just as tobacco and alcohol companies are banned from advertising in publications or programs for children. In general, however, regulating our interactions with technology is a particularly tricky task because degrees of excess depend on individual vulnerabilities.

One case that seems to clearly call for regulation is that of texting while driving. A technology solution should be mandated to stop drivers from texting while their vehicles are in motion. Maybe this could be a granular geofencing of the driver's seat. It also could entail much stricter penalties for accidents caused by texting while driving. But a problem that causes not just loss of attention but loss of lives seems to demand a solution that doesn't wait for an accident to happen. As discussed, Apple introduced an optional feature called "Do Not Disturb While Driving" as part of iOS 11. Whenever the phone is connected to a car using either Bluetooth or a cable, or if the car is moving, the phone can withhold notifications. This raises the question: why shouldn't it be mandatory—and in all mobile phones? No matter that leading technologists, business leaders, and politicians are calling for regulation, it is not in the cards. As a society we still seem to be struggling to come to grips with the reality that we have far less control over our impulses than we want to believe we have. And we are struggling to control our interactions with the most addictive set of technologies ever unleashed. Agency assumes choice, but when choice is dictated by the companies that profit from our addictions and our bad habits, then the road to true reform will be long, bumpy, and painful.

Until the makers find the incentive to limit the push of their technologies into our lives, the best we can do is engineer our lives and our environments to craft a healthier relationship with our tech. We will not always succeed. Changing habits now hard-wired from years of activities is, as any behavioral psychologist can tell you, exceedingly difficult. We have to make these changes, however: for our own sakes, for the sake of society, and for the sake of our children. We might well be the last generation that remembers what life was like before the smartphone so totally dominated our lives as to become truly an extension of our brains and our beings—and likely on terms we have little or no say in setting. If we can build the muscles to wrestle back control and take a clear-eyed view of technologies' effects on our lives, our relationships, and our work, then the future will be much brighter and happier for us, our families, and our communities.

◆ 9 ◆

A Personal Epilogue

Almost immediately after we pitched this book to our publishers, criticism broke out about the business practices, ethics, and values of the big technology companies. In August 2017, sociologist Jean Twenge published her book *iGen*, which examines how teenagers are growing up with technology dominating their lives while being completely unprepared for adulthood.[1] Her September 2017 article in *The Atlantic*, discussed in chapter 6, sparked a firestorm of commentary and criticism. Former *New Republic* editor Franklin Foer published *World Without Mind: The Existential Threat of Big Tech* in September 2017, a polemic that criticizes Google, Facebook, and other tech giants for what he regards as soulless monopolism that seeks to understand every facet of our identities and influence every decision of our lives for profit.[2] In a blog post titled "Hard questions: Is spending time on social media bad for us?" Facebook's director of research, David Ginsberg, finally acknowledged that perhaps the social network was not so good for its users.[3] (The eye-popping irony of the post was that the prescription to solve the problem was even more in-depth Facebook participation!)

That's not all. In early 2018, Roger McNamee stoked

the fires with articles in *Washington Monthly* and the *Washington Post.* Then came an open letter to Tim Cook from Jana Partners and CalSTRS about the impact of the iPhone on children. One of the biggest advocates for technology on the planet then came down hard on Facebook: on January 23, 2018, Salesforce CEO Marc Benioff compared Facebook to the tobacco companies and urged that we regulate the social network.[4]

A backdrop to all this is the steady stream of revelations about how Russian intelligence agencies used the social-media platforms Twitter and Facebook and the search engine Google to target political ads with the aim of swaying the most critical counties in swing states for Donald Trump and away from Hillary Clinton in the 2016 U.S. presidential election. Evidence also emerged that Russia had sought to influence the Brexit vote regarding whether the United Kingdom should leave the European Union.[5]

Within a very short span, we watched the world's suspicions of big technology companies hit a fever pitch. The theme was how technology—specifically products and applications and devices built by the largest, richest technology companies—has come to dominate our lives, stamp out innovation, and put our democratic values at risk. This culminated in October 2017 when the *New York Times* ran a feature article in its Sunday Review opinions and commentary section titled "Silicon Valley is not your friend."[6] The situation had gone from quiet, whispered concerns among our techno-savvy friends to an international debate about how much responsibility the giant tech firms (and all tech firms, really) had for the health and well-being of their users.

This is why, in retrospect, we are doubly glad to be writing this book. However widespread criticisms of tech might be, Big Tech is not going away. Facebook, Google, Twitter, and Amazon are not going to be regulated out of existence. And, frankly, what they do is valuable to us individually and collectively.

It is unlikely that any regulation will happen. That said, the electric cattle prod of nasty public comeuppances has awakened the sleeping giants of Silicon Valley from their slumber and naïveté. They are now at least open to the idea that pushing every psychological button possible to ensure we never leave Facebook or that we buy more and more products from Amazon or that we retweet and post more and more content on Twitter is not only against our best interest but also against the best interest of the companies themselves over the long haul. Something better will come along, with a better monetization model that is more in harmony with the needs of users. Already Google and Facebook are facing higher costs of acquiring users.

Congressional hearings come and go. As observers of Silicon Valley from the pre-Google and pre-Facebook eras, we find that the criticism of technology and its makers, the questioning, resonates with most of the people we have spoken with. A wholesale change in public opinion has occurred. We firmly believe that people—including users—are more closely scrutinizing Big Tech and that we are now readier to push for meaningful changes in the way tech interacts with us than we have been at any time in recent memory.

This is why writing this book has been one of the most rewarding experiences we have ever had. In our discussions with dozens of friends and colleagues, and with thousands of people on line, every single individual we spoke with said that the issues of technology pervasiveness resonate and are of utmost importance to them personally and to society as a whole, showing us how similar their struggles are to our own. They worry about what technology is doing to their brains and to their families. They feel a loss of control over their lives. Some have taken steps to regain their lives. A handful of people aggressively limit the way their children are permitted to use technology. Many people pointed us to primers on how to shut down auto-play videos on Netflix and YouTube. Others have decided to put their phones on Do Not Disturb during the workday.

One senior executive at a Fortune 500 retailer began locking his smartphone in the glove compartment whenever he parked in his driveway. A department at a company where Alex used to work instituted meeting-free Fridays (a policy Alex had championed while he worked there). We spoke to several executives who, at the admin level, shut off employees' e-mail accounts while they were on vacation and set up autoresponders. None of these steps were nuclear strikes against technology. No one stripped away technology use entirely or got off the grid. But many people took steps to regain some control and approach a saner, more balanced life.

These people more than once spoke of beginning to notice all the little ways in which technology had limited their choices and control. Many were saddened by it. They noticed the compulsion to check e-mail while they were out on rare dates with their spouses. Newly sensitized, they began to count all the small ways in which Facebook tries to manipulate them and engage them more deeply. They complained about the difficulty they experienced in putting boundaries on their digital lives across their myriad services, screens, and systems. They were finding it painful and cumbersome to treat the problem.

We also began to slowly change our lives. Alex banned screens in the bedroom and no longer works on computers at home after 6 p.m. (at least, when he is not on book deadlines). He installed a router that is designed to put in place policies on who can use what on the Internet. And in his house, the Internet goes dark at 8 p.m. for everyone. When he needs to work, he goes to a coffee shop or some other place to allow his family a safe space in the evenings, away from technology.

Following the lead of his executive friend, Alex began to shove his phone in the glove box of his car or in the door slot whenever he drives anywhere. He has stopped worrying about making sure that some form of useful information is coming out of the speakers or is on the screen for reading at every moment of the day, freeing him from fiddling with the phone at stop signs to find a replay of a good podcast.

In conversations and in meetings, Alex places his phone

out of sight in his book bag or his pocket. He turns on Do Not Disturb for most of the day, allowing calls to go to voicemail to be triaged later in the day. In writing this book, Alex (like Nicholas Carr) shut off Internet access or wrote in places with no mobile phone or Internet reception. He continues to struggle with a nasty e-mail addiction and may do so for the rest of his life.

Alex has begun using a tool for managing social media, in order to limit his social-media interactions to one check per day. He schedules, in advance, posts of news he finds interesting. He also tries to batch his writing and reading of e-mail into three windows per day. As a result of these changes to his life, he has been able to increase the amount of deep work he is doing, and he now schedules outside time nearly every day: either a walk in the woods or a jog on a trail or some kind of sports with his son and his friends. His kids are very resentful that he now imposes limits on them, but that's not surprising.

Vivek has started turning his computer off for two hours at a time to read and think and take breaks from technology. He turns his computer off and puts his phone into Do Not Disturb mode at 9 p.m. After more than a decade of being hyperactive on social media, Vivek now disconnects from Twitter (his favorite outlet) for a full day each weekend, and for a half a day during the week. He still feels the strong pull to check social media, because he loves the conversations, but he recognizes that they are frequent enough to detract from real-world interactions. He has also stopped responding to the majority of the messages he gets on Twitter, having learned that one usually comes out ahead by not saying anything on social media.

Vivek also schedules regular walks in the county and state parks around Silicon Valley and completely disconnects on Sundays. He has stopped listening to voicemails. And when he's with his children, he puts his devices away.

Like Alex, Vivek continues to struggle with compulsive smartphone and screen behaviors that have been burned into his brain through many years of unconscious consumption. He hopes to further reduce screen time and spend more time reading books and talking with people face to face. He already recognizes that those in-person meetings are both more fulfilling and more useful.

Our goal for this book is that it may actually make an impression on enough people to change their view of technology. We don't want to convince them that technology is dangerous and bad, or that technology should be banned or technological innovation halted. Rather, we want readers to understand the importance of mindful use of technology. All of us should use these products, these miracles of silicon and software, *only on our own terms*. We should demand more control and simpler choices. For their part, the technology companies, and the employers and others who impose technology use upon us or design our user experiences, must begin to factor in the human costs.

And, yes, we do believe that some government intervention may be required. In China, for example, the government has imposed limits on the amount of time per day that children under eighteen can play video games. That doesn't seem like a bad idea here in America. Children are treated differently in every other form of media right now and are prevented from watching movies with excessive violence or nudity. Why not require game makers to limit how long kids and teens can play?

We sincerely hope that all of us have crossed over from the early days of unconscious tech consumption to a period of much more attentive and considered use—and that, in the future, our relationship with our technology will be healthier and more sustainable, without costs we're unwilling to bear. If we can make these kinds of changes, we will all be less lonely and less isolated. We will spend more time talking to each other, walking outside, connecting with nature. We will still have everything technology brings: online maps and navigation, social media, smartphones, e-mail and text, photo sharing, and more. But we will be much smarter about the ways in which we choose to use these tools.

At work, we will have more time to focus on the critical tasks, with fewer interruptions. Our bosses will be happy to see us spend hours of focused work and will not mind that we have not responded to an e-mail promptly. And our vacations will be truly unplugged and restorative. The companies that make all these technological miracles happen will endow us with easier ways to turn off features we don't like and, in general, to design our lives and our days so that tech fits our needs and wishes rather than assuming control over our impulses and lives.

Most important of all, we hope that our children's

children will not be part of a generation unable to conceive of unplugged time or to endure sitting and reading a print book for half an hour.

Yes, we must change the old ways and adapt. We recognize that. But the new ways must not obliterate and dominate the old ones. Those ways are the product of millennia of evolution and clearly have enduring value. Modernity, and its close cousin technology, must not demand that we as users either adapt or be miserable and lonely. Modernity must be a choice that suits us, our children, and our children's children.

NOTES

PREFACE

1. J. Helliwell, R. Layard, and J. Sachs, *World Happiness Report 2017*, New York: Sustainable Development Solutions Network, 2017, http://worldhappiness.report/ed/2017 (accessed 2 February 2018). David G. Blanchflower and Andrew Oswald, *Unhappiness and Pain in Modern America: A Review Essay, and Further Evidence, on Carol Graham's Happiness for All?* NBER Working Paper No. 24087, Cambridge: National Bureau of Economic Research, 2017, http://papers.nber.org/tmp/43780-w24087.pdf (accessed 2 February 2018).

2. Jacqueline Olds and Richard S. Schwartz, *The Lonely American: Drifting Apart in the Twenty-first Century*, Boston: Beacon Press, 2009. American Psychological Association, "So lonely I could die," American Psychological Association news release 5 August 2017, http://www.apa.org/news/press/releases/2017/08/lonely-die.aspx (accessed 2 February 2018).

3. Jean M. Twenge, Thomas E. Joiner, Megan L. Rogers, et al., "Increases in depressive symptoms, suicide-related outcomes, and suicide rates among U.S. adolescents after 2010 and links to increased new media screen time," *Clinical Psychological Science* 2018;6(1):3–17, https://doi.org/10.1177/2167702617723376 (accessed 2 February 2018). Graeme Paton, "Overexposure to technology 'makes children miserable,'" *The Telegraph* 26 October 2012, http://www.telegraph.co.uk/education/educationnews/9636862/Overexposure-to-technology-makes-children-miserable.html (accessed 2 February 2018). Jean M. Twenge, Gabrielle N. Martin, and W. Keith Campbell, "Decreases in psychological well-being among American adolescents after 2012 and links to screen time during the rise of smartphone technology," *Emotion* 2018;18(1), http://dx.doi.org/10.1037/emo0000403 (accessed 2 February 2018).

4. Jean M. Twenge and Heejung Park, "The decline in adult activities among U.S. adolescents, 1976–2016," *Child Development* 2017;00(0):1–17, http://onlinelibrary.wiley.com/doi/10.1111/cdev.12930/ abstract, http://dx.doi.org/10.1111/cdev.12930 (accessed 2 February 2018).

5. "Distracted driving," U.S. Department of Transportation (n.d.), https://www.nhtsa.gov/risky-driving/distracted-driving (accessed 2 February 2018).

6. P. Matthijs Bal and Martijn Veltkamp, "How does fiction reading influence empathy? An experimental investigation on the role of emotional transportation," *PLoS ONE* 2013;8(1):e55341, https://doi .org/10.1371/journal.pone.0055341 (accessed 2 February 2018).

7. Jean M. Twenge, Sara Konrath, Joshua D. Foster, et al., "Egos inflating over time: A cross-temporal meta-analysis of the Narcissistic Personality Inventory," *Journal of Personality* 2008;76(4):875–901, https://doi.org/10.1111/j.1467-6494.2008.00507.x (accessed 2 February 2018). Frederick S. Stinson, Deborah A. Dawson, Risë B. Goldstein, et al., "Prevalence, correlates, disability, and comorbidity of DSM-IV Narcissistic Personality Disorder: Results from the Wave 2 National Epidemiologic Survey on Alcohol and Related Conditions," *Journal of Clinical Psychiatry* 2008;69(7):1033–1045, https://www.psychiatrist .com/jcp/article/Pages/2008/v69n07/v69n0701.aspx (accessed 2 February 2018). Jean M. Twenge and Joshua D. Foster, "Mapping the scale of the narcissism epidemic: Increases in narcissism 2002–2007 within ethnic groups," *Journal of Research in Personality* 2008;42(6):1619–1622, https://doi.org/10.1016/j.jrp.2008.06.014 (accessed 2 February 2018). Jean M. Twenge and Josha D. Foster, "Birth cohort increases in narcissistic personality traits among American college students, 1982–2009," *Social Psychological and Personality Science* 2010;1(1):99–106, https://doi.org/10.1177/1948550609355719 (accessed 2 February 2018). Sara H. Konrath, Edward H. O'Brien, and Courtney Hsing, "Changes in dispositional empathy in American college students over time: A meta-analysis," *Personality and Social Psychology Review* 2010;15(2):180–198. Jean M. Twenge, "The evidence for Generation Me and against Generation We," *Emerging Adulthood* 2013;1(1):11–16, http://journals.sagepub.com/ doi/full/10.1177/2167696812466548 (accessed 2 February 2018). Jean M. Twenge, "Overwhelming evidence for Generation Me: A reply

to Arnett," *Emerging Adulthood* 2013;1(1):21–26, http://journals.sage pub.com/doi/abs/10.1177/2167696812468112 (accessed 2 February 2018). Keith Oatley, "Fiction: Stimulation of social worlds," *Trends in Cognitive Sciences* 2016;20(8):618–628, http://dx.doi.org/10.1016/ j.tics.2016.06.002 (accessed 2 February 2018). Joe Pierre, "The narcissism epidemic and what we can do about it: Looking in the mirror at our love of narcissists, part 3," *Psychology Today* 8 July 2016, https:// www.psychologytoday.com/blog/psych-unseen/201607/the-narcissism -epidemic-and-what-we-can-do-about-it (accessed 2 February 2018).

8. *Trends in Consumer Mobility Report 2015*, Bank of America 2015, http://newsroom.bankofamerica.com/files/doc_library/additional/ 2015_BAC_Trends_in_Consumer_Mobility_Report.pdf (accessed 2 February 2018).

9. Timothy D. Wilson, David A. Reinhard, Erin C. Westgate, et al., "Just think: The challenges of the disengaged mind," *Science* 2014;345(6192):75–77, http://dx.doi.org/10.1126/science.1250830 (accessed 2 February 2018).

10. Paul Lewis, "'Our minds can be hijacked': The tech insiders who fear a smartphone dystopia," *The Guardian* 6 October 2017, http:// www.theguardian.com/technology/2017/oct/05/smartphone-addiction -silicon-valley-dystopia (accessed 2 February 2018).

11. Scott Berinato, "Inside Facebook's A.I. workshop," *Harvard Business Review* 19 July 2017, https://hbr.org/2017/07/inside-facebooks -ai-workshop (accessed 2 February 2018).

12. Jacqueline Howard, "Americans devote more than 10 hours a day to screen time, and growing," CNN 29 July 2016, https://edition .cnn.com/2016/06/30/health/americans-screen-time-nielsen/index .html (accessed 2 February 2018). Sarah Perez, "U.S. consumers now spend 5 hours per day on mobile devices," *TechCrunch* 3 March 2017, https://techcrunch.com/2017/03/03/u-s-consumers-now-spend-5 -hours-per-day-on-mobile-devices (accessed 2 February 2018).

INTRODUCTION

1. Noah Budnick, "Largest distracted driving behavior study," Zendrive 17 April, http://blog.zendrive.com/distracted-driving (accessed 2 February 2018).

2. David Comer Kidd and Emanuele Castano, "Reading literary

fiction improves theory of mind," *Science* 2013;342(6156):377–380, https://dx.doi.org/10.1126/science.1239918 (accessed 2 February 2018).

3. Victoria Rideout, *Children, Teens, and Reading: A Common Sense Media Research Brief*, San Francisco: Common Sense Media, 2014, https://www.commonsensemedia.org/file/csm-childrenteensandread ing-2014pdf/download (accessed 2 February 2018). American Academy of Arts and Sciences, "Youth reading for fun," Humanities Indicators January 2016 (accessed 2 February 2018), https://humanitiesindica tors.org/content/indicatordoc.aspx?i=10975 (accessed 2 February 2018). "Results from the Annual Arts Basic Survey (2013–2015)," National Endowment for the Arts August 2016, https://www.arts.gov/ artistic-fields/research-analysis/arts-data-profiles/arts-data-profile -10 (accessed 2 February 2018). Margaret K. Merga and Saiyidi Mat Roni, "The influence of access to eReaders, computers and mobile phones on children's book reading frequency," *Computers & Education* 2017;109:187–196, https://doi.org/10.1016/j.compedu.2017.02.016 (accessed 2 February 2018).

4. Naomi S. Baron, *Words Onscreen: The Fate of Reading in a Digital World*, Oxford: OUP, 2015, https://global.oup.com/academic/ product/words-onscreen-9780199315765 (accessed 2 February 2018). Anne Mangen, Bente R. Walgermo, and Kolbjørn Brønnick, "Reading linear texts on paper versus computer screen: Effects on reading com- prehension," *International Journal of Educational Research* 2013;58:61– 68, https://dx.doi.org/10.1016/j.ijer.2012.12.002 (accessed 2 February 2018).

5. Perry W. Thorndyke and Barbara Hayes-Roth, "Differences in spatial knowledge acquired from maps and navigation," *Cognitive Psychology* 1982;14(4):560–589, https://pdfs.semanticscholar.org/ e847/34d4504d7b85db0500b8409c72ce26b29160.pdf (accessed 2 February 2018). Ginette Wessel, Caroline Ziemkiewicz, Remco Chang, et al., *GPS and road map navigation: The case for a spatial framework for semantic information*, Rome: Advanced Visual Interfaces '10 conference presentation, 2010. Lin Edwards, "Study suggests reliance on GPS may reduce hippocampus function as we age," *Medical Xpress* 18 November 2010, https://medicalxpress.com/news/2010-11-reliance-gps-hippo campus-function-age.html (accessed 2 February 2018). Katherine Woollett and Eleanor A. Maguire, "Acquiring 'the Knowledge' of London's layout drives structural brain changes," *Current Biology*

2011;21(24):2109–2114, https://doi.org/10.1016/j.cub.2011.11.018. Stefan Münzer, Hubert D. Zimmer, and Jörg Baus, "Navigation assistance: A trade-off between wayfinding support and configural learning support," *Journal of Experiential Psychology Applied* 2012;18(1):18–37, https://dx.doi.org/10.1037/a0026553 (accessed 2 February 2018). Toru Ishikawa and Kazunori Takahashi, "Relationships between methods for presenting information on navigation tools and users' wayfinding behavior," *Cartographic Perspectives* 2013;(75):17–28, https://dx.doi.org/ 10.14714/CP75.82 (accessed 2 February 2018). John Edward Huth, *The Lost Art of Finding Our Way*, Cambridge: Harvard University Press, 2013. Amir-Homayoun Javadi, Beatrix Emo, Lorelei R. Howard, et al., "Hippocampal and prefrontal processing of network topology to simulate the future," *Nature Communications* 2017;8:14652, https://dx.doi .org/10.1038/ncomms14652 (accessed 2 February 2018). Leon Neyfakh, "Do our brains pay a price for GPS?" *The Boston Globe* 17 August 2013, https://www.bostonglobe.com/ideas/2013/08/17/our-brains-pay-price -for-gps/d2Tnvo4hiWjuybid5UhQVO/story.html (accessed 2 February 2018).

6. Daniel Kahneman, *Thinking, Fast and Slow*, New York: Farrar, Straus and Giroux, 2011.

7. John M. Jakicic, Kelliann K. Davis, Renee J. Rogers, et al., "Effect of wearable technology combined with a lifestyle intervention on long-term weight loss: The IDEA randomized clinical trial," *JAMA* 2016;316(11):1161–1171, http://dx.doi.org/10.1001/jama.2016.12858 (accessed 2 February 2018).

8. Courtney C. Simpson and Suzanne E. Mazzeo, "Calorie counting and fitness tracking technology: Associations with eating disorder symptomatology," *Eating Behaviours* 2017;26:89–92, doi.org/10.1016/ j.eatbeh.2017.02.002 (accessed 2 February 2018).

9. Thuy Ong, "39 million Americans reportedly own a voice-activated smart speaker," *The Verge* 15 Jan 2018, https://www.theverge .com/2018/1/15/16892254/smart-speaker-ownership-google-amazon (accessed 2 February 2018).

10. Dana Hull, "Elon Musk's Neuralink gets $27 million to build brain computers," *Bloomberg Technology* 25 Aug 2017, https://www .bloomberg.com/news/articles/2017-08-25/elon-musk-s-neuralink-gets -27-million-to-build-brain-computers (accessed 2 February 2018).

11. Vivek Wadhwa and Alex Salkever, *The Driver in the Driverless*

Car: How Our Technology Choices Will Create the Future, Oakland, California: Berrett-Koehler, 2017.

CHAPTER I

1. Jessica Lee, "No. 1 position in Google gets 33% of search traffic [study]," Search Engine Watch 11 February 2018, https://searchengine watch.com/sew/study/2276184/no-1-position-in-google-gets-33-of -search-traffic-study (accessed 2 February 2018).

2. "The way the brain buys," *The Economist* 18 December 2008, http://www.economist.com/node/12792420 (accessed 2 February 2018).

3. Tristan Harris, "How technology is hijacking your mind—from a magician and Google design ethicist," *Thrive Global* 18 May 2016, journal.thriveglobal.com/how-technology-hijacks-peoples-minds -from-a-magician-and-google-s-design-ethicist-56d62ef5edf3 (accessed 2 February 2018).

4. Charles B. Ferster and B. F. Skinner, *Schedules of Reinforcement*, New York: Appleton-Century-Crofts, 1957.

5. Andrew Thompson, "Engineers of addiction," *The Verge* 6 May 2015, https://www.theverge.com/2015/5/6/8544303/casino-slot -machine-gambling-addiction-psychology-mobile-games (accessed 2 February 2018).

6. Brad Plumer, "Slot-machine science: How casinos get you to spend more money," *Vox* 1 Mar 2015, https://www.vox.com/2014/8/ 7/5976927/slot-machines-casinos-addiction-by-design (accessed 2 February 2018).

7. Candice Graydon, Mike J. Dixon, Kevin A. Harrigan, et al., "Losses disguised as wins in multiline slots: using an educational animation to reduce erroneous win overestimates," *International Gambling Studies* 2017;17:442–458, http://www.tandfonline.com/doi/ full/10.1080/14459795.2017.1355404 (accessed 2 February 2018).

8. Luke Clark, Andrew J. Lawrence, Frances Astley-Jones, et al., "Gambling near-misses enhance motivation to gamble and recruit win-related brain circuitry," *Neuron* 2009;61(3):481–490, https://doi .org/10.1016/j.neuron.2008.12.031 (accessed 2 February 2018).

9. Mark R. Dixon and Jacob Daar, "Losses disguised as wins, the science behind casino profits," *The Conversation* 3 November 2014,

https://theconversation.com/losses-disguised-as-wins-the-science
-behind-casino-profits-31939 (accessed 2 February 2018).

10. Robert B. Breen and Mark Zimmerman, "Rapid onset of
pathological gambling in machine gamblers," *Journal of Gambling
Studies* 2002;18(1):31–43, https://doi.org/10.1023/A:1014580112648
(accessed 2 February 2018). Mike J. Dixon, Kevin A. Harrigan,
Rajwant Sandhu, et al., "Losses disguised as wins in modern multiline video slot machines," *Addiction* 2010;105(10):1819–1824, https://
dx.doi.org/10.1111/j.1360-0443.2010.03050.x (accessed 2 February
2018). "Congratulations, you've lost! How slot machines disguise
losses as wins," *Freakonomics* 1 September 2011, http://freakonomics
.com/2011/09/01/congratulations-youve-lost-how-slot-machines-dis
guise-loses-as-wins (accessed 2 February 2018). Alice Robb, "Why are
slot machines so addictive?" *New Republic* 5 December 2013, https://
newrepublic.com/article/115838/gambling-addiction-why-are-slot
-machines-so-addictive (accessed 2 February 2018). Brad Plumer, "Slot-
machine science: How casinos get you to spend more money," *Vox*
1 Mar 2015, https://www.vox.com/2014/8/7/5976927/slot-machines
-casinos-addiction-by-design (accessed 2 February 2018). Candice
Graydon, Mike J. Dixon, Kevin A. Harrigan, et al., "Losses disguised
as wins in multiline slots: Using an educational animation to reduce
erroneous win overestimates." K. R. Barton, Y. Yazdani, N. Ayer, et al.,
"The effect of losses disguised as wins and near misses in electronic
gaming machines: A systematic review," *Journal of Gambling Studies*
2017;33:1241–1260, https://doi.org/10.1007/s10899-017-9688-0 (accessed 2 February 2018).

11. Tristan Harris, "How technology is hijacking your mind—
from a magician and Google design ethicist."

12. Keith Hampton, Lauren Sessions Goulet, Eun Ja Her, et
al., *Social Isolation and New Technology: How the Internet and Mobile
Phones Impact Americans' Social Networks*, Washington, DC: Pew
Research Center, 2009, http://www.pewinternet.org/2009/11/04/
social-isolation-and-new-technology (accessed 2 February 2018).

13. David Ginsbert and Moira Burke, "Hard questions: Is spending time on social media bad for us?" Facebook Newsroom 15 Dec
2017, https://newsroom.fb.com/news/2017/12/hard-questions-is
-spending-time-on-social-media-bad-for-us (accessed 2 February 2018).

14. Eli J. Finkel, Paul W. Eastwick, Benjamin R. Karney, et al.,

"Online dating: A critical analysis from the perspective of psychological science," *Psychological Science in the Public Interest* 2012;13(1):3–66, http://www.psychologicalscience.org/publications/journals/pspi/online-dating.html (accessed 2 February 2018). Hui-Tzu Grace Chou and Nicholas Edge, "'They are happier and having better lives than I am': The impact of using Facebook on perceptions of others' lives," *Cyberpsychology, Behavior, and Social Networking* 2012;15(2):117–121, https://doi.org/10.1089/cyber.2011.0324 (accessed 2 February 2018). Sonja Lyubomirsky and Lee Ross, "Hedonic consequences of social comparison: A contrast of happy and unhappy people," *Journal of Personality and Social Psychology* 1997;73(6):1141–1157, https://www.ncbi.nlm.nih.gov/pubmed/9418274 (accessed 2 February 2018).

15. Emily Hanna, L. Monique Ward, Rita C. Seabrook, et al., "Contributions of social comparison and self-objectification in mediating associations between Facebook use and emergent adults' Psychological Well-Being," *Cyberpsychology, Behavior, and Social Networking* 2017;20(3)172–179, https://doi.org/10.1089/cyber.2016.0247 (accessed 2 February 2018). Helmut Appel, Alexander L. Gerlach, and Jan Crusius, "The interplay between Facebook use, social comparison, envy, and depression," *Current Opinion in Psychology* 2016 June;9:44–49, https://doi.org/10.1016/j.copsyc.2015.10.006 (accessed 2 February 2018).

16. Allee Manning, "Teens are crippled by social media-fueled FOMO," *Vocativ* 16 June 2016, http://www.vocativ.com/329926/teen-social-media-fomo/index.html (accessed 2 February 2018).

17. Sebastián Valenzuela, Daniel Halpern, and James E.Katz, "Social network sites, marriage well-being and divorce: Survey and state-level evidence from the United States," *Computers in Human Behavior* 2014;36:94–101, https://doi.org/10.1016/j.chb.2014.03.034 (accessed 2 February 2018).

18. Emily Hanna, L. Monique Ward, Rita C. Seabrook, et al., "Contributions of social comparison and self-objectification in mediating associations between Facebook use and emergent adults' psychological well-being," *Cyberpsychology, Behavior, and Social Networking* 2017;20(3):172–179, https://doi.org/10.1089/cyber.2016.0247 (accessed 2 February 2018).

19. Holly B. Shakya and Nicholas A. Christakis, "Association of Facebook use with compromised well-being: A longitudinal study,"

American Journal of Epidemiology 2017;185(3):203–211, https://doi.org/10.1093/aje/kww189 (accessed 2 February 2018).

20. Proma Khosla, "Study reveals how often we laugh, cry, and get creeped on while watching Netflix in public," Mashable Australia 14 November 2017, https://mashable.com/2017/11/14/netflix-public-bingeing (accessed 22 March 2018).

21. Virginia Lau, "The Michael J. Fox Foundation uses Facebook to recruit Ashkenazi Jews for Parkinson's study," *MM&M* 22 Sep 2016, http://www.mmm-online.com/campaigns/the-michael-j-fox-foundation-uses-facebook-to-recruit-ashkenazi-jews-for-parkinsons-study/article/524267 (accessed 2 February 2018).

22. Liese Exelmans and Jan Van den Buick, "Binge viewing, sleep, and the role of pre-sleep arousal," *Journal of Clinical Sleep Medicine* 2017;13(8):1001–1008, https://dx.doi.org/10.5664/jcsm.6704 (accessed 2 February 2018).

23. Alex Hern, "Netflix's biggest competitor? Sleep," *The Guardian* 18 April 2017, https://www.theguardian.com/technology/2017/apr/18/netflix-competitor-sleep-uber-facebook (accessed 2 February 2018).

24. Gloria Mark, Shamsi T. Iqbal, Mary Czerwinski, et al., "Email duration, batching and self-interruption: Patterns of email use on productivity and stress," in *Proceedings of the 2016 CHI Conference on Human Factors in Computing Systems*, New York: ACM, 2016, http://dx.doi.org/10.1145/2858036.2858262 (accessed 2 February 2018).

25. Andrew Nusca, "Slack raises $250 million; tops $5 billion variation," *Fortune* 18 September 2017, http://fortune.com/2017/09/17/slack-raise-valuation (accessed 2 February 2018).

26. Gloria Mark, Shamsi T. Iqbal, Mary Czerwinski, et al., "Email duration, batching and self-interruption: Patterns of email use on productivity and stress."

27. Tristan Harris, "How technology is hijacking your mind—From a magician and Google design ethicist."

28. Jenny Anderson, "A letter from two big Apple investors powerfully summarizes how smartphones mess with kids' brains," *Quartz* 8 Jan 2018, https://qz.com/1174317/a-letter-from-apple-aapl-investors-jana-partners-and-calstrs-powerfully-summarizes-how-smartphones-mess-with-kids-brains (accessed 2 February 2018).

29. "IAB ad blocking report: Who blocks ads, why, and how to

win them back," IAB 26 July 2016, https://www.iab.com/insights/ad
-blocking-blocks-ads-win-back (accessed 2 February 2018).

30. Sally Andrews, David A. Ellis, Heather Shaw, et al., "Beyond
self-report: Tools to compare estimated and real-world smartphone
use," *PLoS ONE* 2015;10(10):e0139004, https://doi.org/10.1371/journal
.pone.0139004 (accessed 2 February 2018).

31. Glen Fleishman, "'Stranger Danger' to children vastly over-
stated," *BoingBoing* 24 Feb 2015, https://boingboing.net/2015/02/24/
our-children-are-safer-than-ou.html (accessed 2 February 2018).

32. Hanna Rosin, "The overprotected kid," *The Atlantic* April
2014, https://www.theatlantic.com/magazine/archive/2014/04/
hey-parents-leave-those-kids-alone/358631 (accessed 2 February
2018). Victoria Rideout, *Children, Teens, and Reading*, San Francisco:
Common Sense Media, 2014, https://www.commonsensemedia.org/
file/csm-childrenteensandreading-2014pdf/download (accessed 2
February 2018). *News and America's Kids: How Young People Perceive
and Are Impacted by the News*, San Francisco: Common Sense Media,
2017, https://www.commonsensemedia.org/research/news-and-ameri
cas-kids (accessed 2 February 2018).

33. Ashley J. Thomas, P. Kyle Stanford, and Barbara W. Sarnecka,
"No child left alone: Moral judgments about parents affect estimates
of risk to children," *Collabra* 2016;2(1):10, http://doi.org/10.1525/
collabra.33 (accessed 2 February 2018). Ashley J. Thomas, P. Kyle
Stanford, and Barbara W. Sarnecka, "Correction: No child left alone:
Moral judgments about parents affect estimates of risk to children,"
Collabra 2016;2(1):12, http://doi.org/10.1525/collabra.58 (accessed 2
February 2018).

34. Melissa L. Finucane, Ali Alhakami, Paul Slovic, et al.,
"The affect heuristic in judgements of risks and benefits," *Journal
of Behavioral Decision Making* 2000;13:1–17, http://www.anderson
.ucla.edu/faculty/keith.chen/negot. papers/FinAlhSlovicJohn_Affect
Heuro0.pdf (accessed 2 February 2018).

35. Carmen Keller, Michael Siegrist, and Heinz Gutscher, "The
role of the affect and availability heuristics in risk communication,"
Risk Analysis 2006;26(3):631–639, https://dx.doi.org/10.1111/j.1539
-6924.2006.00773.x (accessed 2 February 2018).

36. Marc Andrews, Matthijs van Leeuwen, and Rick van Baaren,
Hidden Persuasion: 33 Psychological Influence Techniques in Advertising,

Amsterdam: BIS, 2013. Natasha Dow Schüll, *Addiction by Design: Machine Gambling in Las Vegas,* Princeton: Princeton University Press, 2014.

CHAPTER 2

1. Ian Leslie, "The scientists who make apps addictive," *1843* October–November 2016, https://www.1843magazine.com/features/the-scientists-who-make-apps-addictive (accessed 2 February 2018).
2. Ibid.
3. Nir Eyal, *Hooked: How to Build Habit-Forming Products,* New York: Penguin, 2014.
4. "Habit Summit," Habit Summit (n.d.), https://habitsummit.com (accessed 2 February 2018).
5. B. J. Fogg and Clifford Nass, "How users reciprocate to computers: An experiment that demonstrates behavior change," CHI 97 Electronic Publications: Late-Breaking/Short Talks, https://archive.is/uvuzh (accessed 2 February 2018).
6. "What is captology," Stanford Persuasive Technology Lab (n.d.), http://captology.stanford.edu/about/what-is-captology.html (accessed 2 February 2018).

CHAPTER 3

1. Renee Stepler, "Led by baby boomers, divorce rates climb for America's 50+ population," Pew Research Center 9 March 2017, http://www.pewresearch.org/fact-tank/2017/03/09/led-by-baby-boomers-divorce-rates-climb-for-americas-50-population (accessed 2 February 2018). "National marriage and divorce rate trends," CDC National Center for Health Statistics, 23 November 2015, https://www.cdc.gov/nchs/nvss/marriage_divorce_tables.htm (accessed 2 February 2018).
2. Richard Fry, "New census data show more Americans are tying the knot, but mostly it's the college-educated," Pew Research Center 6 February 2014, http://www.pewresearch.org/fact-tank/2014/02/06/new-census-data-show-more-americans-are-tying-the-knot-but-mostly-its-the-college-educated (accessed 2 February 2018).
3. Jean M. Twenge, Ryne A. Sherman, and Brooke E. Wells, "Declines in sexual frequency among American adults, 1989–2014,"

Archives of Sexual Behavior 2017;46(8):2389–2401, https://dx.doi.org/
10.1007/s10508-017-0953-1 (accessed 2 February 2018).

4. John F. Helliwell and Shawn Grover, *How's Life at Home? New Evidence on Marriage and the Set Point for Happiness* (NBER Working Paper 20794), Cambridge: National Bureau of Economic Research, 2014, https://www.nber.org/papers/w20794, https://dx.doi.org/10 .3386/w20794 (accessed 2 February 2018).

5. Rebecca Rosen, "Marriage will not fix poverty," *The Atlantic* 11 March 2016, http://www.theatlantic.com/business/archive/2016/03/ marriage-poverty/473019 (accessed 2 February 2018).

6. Jane Anderson, "The impact of family structure on the health of children: Effects of divorce," *The Linacre Quarterly* 2014;81(4):378– 387, https://doi.org/10.1179/0024363914Z.00000000087 (accessed 2 February 2018).

7. Emerging Technology, "First evidence that online dating is changing the nature of society," *MIT Technology Review* 10 October 2017, http://www.technologyreview.com/s/609091/first-evidence-that -online-dating-is-changing-the-nature-of-society (accessed 2 February 2018).

8. Aaron Smith, *15% of American Adults Have Used Online Dating Sites or Mobile Dating Apps*, Washington, DC: Pew Research Center, 2016, http://www.pewinternet.org/2016/02/11/15-percent-of-american -adults-have-used-online-dating-sites-or-mobile-dating-apps (accessed 2 February 2018).

9. Catalina L. Toma, Jeffrey T. Hancock, and Nicole B. Ellison, "Separating fact from fiction: An examination of deceptive self- presentation in online dating profiles," *Personality and Social Psychology Bulletin* 2008;34(8):1023–1036, https://doi.org/10.1177/ 0146167208318067 (accessed 2 February 2018).

10. Marisa Meltzer, "Online dating: Match me if you can," *Consumer Reports* 29 December 2016, www.consumerreports.org/dat ing-relationships/online-dating-guide-match-me-if-you-can (accessed 2 February 2018).

11. John T. Cacioppo et al., "Marital satisfaction and break-ups differ across on-line and off-line meeting venues," *Proceedings of the National Academy of Sciences USA* 2013;110(25):10135–10140, https:// doi.org/10.1073/pnas.1222447110 (accessed 2 February 2018).

12. Amy Webb, *Data: A Love Story*, New York: Penguin, 2013.

13. Jean M. Twenge, Ryne A. Sherman, and Brooke E. Wells, "Changes in American adults' sexual behavior and attitudes, 1972–2012," *Archives of Sexual Behavior* 2015;44(8):2273–2285, https://doi .org/10.1007/s10508-015-0540-2 (accessed 2 February 2018).

14. Eli J. Finkel, Paul W. Eastwick, Benjamin R. Karney, et al., "Online dating: A critical analysis from the perspective of psychological science," *Psychological Science in the Public Interest* 2012;3(1):3–66, https://doi.org/10.1177/1529100612436522 (accessed 2 February 2018).

15. Eli J. Finkel, Paul W. Eastwick, Benjamin R. Karney, et al., "Dating in a digital world," *Scientific American Mind* 2012;23:26–33, https://www.nature.com/scientificamericanmind/journal/v23/n4/full, https://dx.doi.org/10.1038/scientificamericanmind0912-26, http:// www.quizner.co/OnlineDating.pdf (accessed 2 February 2018).

16. Maris Kreizmen, "An algorithm isn't always the answer," *New York Times* 24 November 2017, https://www.nytimes.com/2017/11/24/ opinion/sunday/holidays-gifts-algorithms-online-dating.html (accessed 2 February 2018).

17. Eli J. Finkel, Paul W. Eastwick, Benjamin R. Karney, et al., "Dating in a digital world."

18. Eli J. Finkel, Paul W. Eastwick, Benjamin R. Karney, et al., "Online dating: A critical analysis from the perspective of psychological science."

19. Jeffrey A. Hall and Benjamin L. Compton, "Pre- and post interaction physical attractiveness ratings and experience-based impressions," *J Communication Studies* 2017;68(3):260–277, https://dx.doi .org/10.1080/10510974.2017.1317281 (accessed 2 February 2018).

20. Ibid.

21. Eli J. Finkel, Paul W. Eastwick, Benjamin R. Karney, et al., "Dating in a digital world."

22. Eli J. Finkel, Paul W. Eastwick, Benjamin R. Karney, et al., "Online dating: A critical analysis from the perspective of psychological science."

23. "We Experiment on Human Beings!," OkCupid 27 July 2014, theblog.okcupid.com/we-experiment-on-human-beings -5dd9fe28ocd5.

24. Jonathan D. D'Angelo and Catalina L. Toma, "There are plenty of fish in the sea: The efects of choice overload and reversibility on online daters' satisfaction with selected partners," *Media Psychology*

2017;20(1):1–27, https://doi.org/10.1080/15213269.2015.1121827 (accessed 2 February 2018).

25. Jennie Zhang and Taha Yasseri, *What Happens After You Both Swipe Right: A Statistical Description of Mobile Dating Communications*, Oxford: Oxford Internet Institute, 2016, arxiv.org/pdf/1607.03320.pdf (accessed 2 February 2018).

26. Julie Beck, "The rise of dating-app fatigue," *The Atlantic* 25 October 2016, http://www.theatlantic.com/health/archive/2016/10/the-unbearable-exhaustion-of-dating-apps/505184 (accessed 2 February 2018).

27. Jessica Strubel and Trent A. Petrie, "Love me Tinder: Body image and psychosocial functioning among men and women," *Body Image* 2017;21:34–38, https://doi.org/10.1016/j.bodyim.2017.02.006 (accessed 2 February 2018).

28. "2017 Year in Review," Pornhub Insights 9 Jan 2018, https://www.pornhub.com/insights/2017-year-in-review (accessed 2 February 2018).

29. Kirsten Weir, "Is pornography addictive?" *American Psychological Association* 2014 April;45(4):46, https://www.apa.org/monitor/2014/04/pornography.aspx (accessed 2 February 2018).

30. Paul J. Wright, Ana Bridges, Chyng Sun, et al., "Personal pornography viewing and sexual satisfaction: A quadratic analysis," *Journal of Sex & Marital Therapy* 2017;8:1–8, https://dx.doi.org/10.1080/0092623X.2017.1377131.

31. Jenny Minarcik, "The effects of sexually explicit material use on romantic relationship dynamics," *Journal of Behavioral Addictions* 2016;5(4):700–707, https://dx.doi.org/10.1556/2006.5.2016.078 (accessed 2 February 2018).

32. Nathaniel M. Lambert, "A love that doesn't last: Pornography consumption and weakened commitment to one's romantic partner," *Journal of Social and Clinical Psychology* 2012;31(4):410–438.

33. Simone Kühn and Jürgen Gallinat, "Brain structure and functional connectivity associated with pornography consumption," *JAMA Psychiatry* 2014;71(7):827–834, https://dx.doi.org/10.1001/jamapsychiatry.2014.93(accessed 2 February 2018).

34. Jason S. Carroll and Brian J. Willoughby, "The porn gap: Differences in men's and women's pornography patterns in couple relationships," Institute for Family Studies 4 October 2017, https://

ifstudies.org/blog/the-porn-gap-gender-differences-in-pornography
-use-in-couple-relationships (accessed 2 February 2018).

35. J. Michael Bostwick and Jeffrey A. Bucci, "Internet sex addiction treated with naltrexone," *Mayo Clinic Proceedings* 2008;83(2): 226–230, http://dx.doi.org/10.4065/83.2.226 (accessed 2 February 2018).

36. Frank O. Poulsen, Dean M. Busby, and Adam M. Galovan, "Pornography use: Who uses it and how it is associated with couple outcomes," *Journal of Sex Research* 2013;50(1):72–83, http://www.tand fonline.com/doi/abs/10.1080/00224499.2011.648027 (accessed 2 February 2018).

37. Jean M. Twenge, Ryne A. Sherman, and Brooke E. Wells, "Declines in sexual frequency among American adults, 1989–2014."

38. Christopher Ryan, "The Future of Sex," *Psychology Today* 5 May 2014, http://www.psychologytoday.com/blog/sex-dawn/201405/ the-future-sex (accessed 2 February 2018).

39. Amy Mulse, Ulrich Schimmack, and Emily A. Impett, "Sexual frequency predicts greater well-being, but more is not always better," *Social Psychological and Personality Science* 2016;7(4):295–302, http:// journals.sagepub.com/doi/abs/10.1177/1948550615616462 (accessed 2 February 2018). Howard S. Friedman, "Orgasms, health and longevity: Does sex promote health?" *Psychology Today* 12 February 2011, https:// www.psychologytoday.com/blog/secrets-longevity/201102/orgasms -health-and-longevity-does-sex-promote-health (accessed 2 February 2018).

40. Anik Debrot, Nathalie Meuwly, Amy Muise, et al., "More than just sex," *Personality and Social Psychology Bulletin* 2017;43(3):287–299, https://doi.org/10.1177/0146167216684124 (accessed 2 February 2018).

CHAPTER 4

1. Cal Newport, *Deep Work: Rules for Focused Success in a Distracted World*, New York: Grand Central Publishing, 2016.

2. Nicholas Carr, *The Shallows: What the Internet Is Doing to Our Brains*, New York: W. W. Norton, 2011.

3. Jeffrey Pfeffer, *Dying for a Paycheck: How Modern Management Harms Employee Health and Company Performance—and What We Can Do About It*, New York: HarperBusiness, 2018.

4. Megan Gibson, "Here's a radical way to end vacation email overload," *Time* 15 August 2014, http://time.com/3116424/daimler -vacation-email-out-of-office (accessed 2 February 2018).

5. Chris Weller, "10 companies that pay employees extra to take vacations," *Inc.* 3 October 2017, http://www.inc.com/business-insider/ 10-companies-that-pay-employees-extra-for-vacations.html (accessed 2 February 2018).

6. Informate, "No time to talk: Americans sending/receiving five times as many texts compared to phone calls each day, according to new report," PR Newswire news release 25 March 2015, https://www .prnewswire.com/news-releases/no-time-to-talk-americans-sending receiving-five-times-as-many-texts-compared-to-phone-calls-each-day -according-to-new-report-300056023.html (accessed 2 February 2018).

7. Michael Malkins, "Is technology really helping us get more done?" *Harvard Business Review* 25 February 2016, hbr.org/2016/02/ is-technology-really-helping-us-get-more-done (accessed 2 February 2018).

8. *Email Statistics Report, 2017–2021*, London: The Radicati Group, 2017, https://www.radicati.com/?p=14588 (accessed 2 February 2018).

9. "Connecting one billion users every day," WhatsApp Blog 26 July 2017, blog.whatsapp.com/10000631/Connecting-One-Billion -Users-Every-Day (accessed 2 February 2018).

10. *Instant Messaging Market, 2017–2021*, London: The Radicati Group, 2017, https://www.radicati.com/?p=14611 (accessed 2 February 2018).

11. Gloria Mark, Daniela Gudith, and Ulrich Klocke, "The Cost of Interrupted Work: More Speed and Stress," in *Proceedings of the 2008 Conference on Human Factors in Computing Systems, CHI 2008*, New York: ACM Press, 2008, https://dx.doi.org/10.1145/1357054.1357072 (accessed 2 February 2018).

12. Gloria Mark, Shamsi T. Iqbal, Mary Czerwinski, et al., "Neurotics can't focus: An *in situ* study of online multitasking in the workplace," in *Proceedings of the 2016 CHI Conference on Human Factors in Computing Systems*, New York: ACM Press, 2016, https://doi .org/10.1145/2858036.2858202 (accessed 2 February 2018).

13. Kermit Pattison, "Worker, interrupted: The cost of task switching," *Fast Company* 28 July 2008, https://www.fastcompany.com/

944128/worker-interrupted-cost-task-switching (accessed 2 February 2018).

14. Mary Meeker and Liang Wu, "Internet trends report 2013," Slideshare 29 May 2013, https://www.slideshare.net/kleinerperkins/kpcb-internet-trends-2013/52-Mobile_Users_Reach_to_Phone (accessed 2 February 2018).

15. Sophie Leroy, "Why is it so hard to do my work? The challenge of attention residue when switching between work tasks," *Organizational Behavior and Human Decision Processes* 2009;109(2)168–181, https://doi.org/10.1016/j.obhdp.2009.04.002 (accessed 2 February 2018).

16. Schumpeter, "The collaboration curse: The fashion for making employees collaborate has gone too far," *The Economist*, 23 January 2016, http://www.economist.com/news/business/21688872-fashion-making-employees-collaborate-has-gone-too-far-collaboration-curse (accessed 2 February 2018).

17. Susan Payne Carter, Kyle Greenberg, and Michael S. Walker, "The impact of computer usage on academic performance: Evidence from a randomized trial at the United States Military Academy," *Economics of Education Review* 2017;56:118–132, https://doi.org/10.1016/j.econedurev.2016.12.005 (accessed 2 February 2018).

18. Dan Nixon, "Is the economy suffering a crisis of attention?" *Bank Underground* 24 November 2017, https://bankunderground.co.uk/2017/11/24/is-the-economy-suffering-from-the-crisis-of-attention (accessed 2 February 2018).

19. Gloria Mark, Daniela Gudith, and Ulrich Klocke, "The cost of interrupted work: More speed and stress."

20. "5 questions with Gloria Mark," Aspen Ideas Festival 22 June 2016, http://www.aspenideas.org/blog/5-questions-gloria-mark (accessed 2 February 2018).

21. Tom Monahan, "The hard evidence: Business is slowing down," *Fortune* 28 January 2016, http://fortune.com/2016/01/28/business-decision-making-project-management (accessed 2 February 2018).

22. Michael Mankins, "Is technology really helping us get more done?"

23. Ibid.

24. Michael Mankins, "This weekly meeting took up 300,000

hours a year," *Harvard Business Review* 29 April 2014, https://hbr.org/
2014/04/how-a-weekly-meeting-took-up-300000-hours-a-yearOther
(accessed 2 February 2018).

25. Catherine Clifford, "How much time do your employees
spend doing real work? The answer may surprise you," *Entrepreneur*
23 November 2014, https://www.entrepreneur.com/article/240076
(accessed 2 February 2018).

26. "How technology impacts workplace productivity today (info-
graphic)," *Highfive* 20 August 2015, highfive.com/blog/impact-of-tech
nology-on-productivity (accessed 2 February 2018).

27. M. Mahdi Roghanizad and Vanessa K. Bohns, "Ask in
person: You're less persuasive than you think over email," *Journal of
Experimental Social Psychology* 2017;69:223–226, https://doi.org/10
.1016/j.jesp.2016.10.002 (accessed 2 February 2018).

28. Sarah Kessler, "IBM, remote-work pioneer, is calling thousands of employees back to the office," *Quartz* 21 March 2017, qz.com/
924167/ibm-remote-work-pioneer-is-calling-thousands-of-employees
-back-to-the-office (accessed 2 February 2018).

29. Wayne F. Cascio and Ramiro Montealegre, "How technology
is changing work and organizations," *Annual Review of Organizational
Psychology and Organizational Behavior* 3:349–375, https://doi.org/10
.1146/annurev-orgpsych-041015-062352 (accessed 2 February 2018).

30. Sarah Kessler, "Researchers have settled the question of
whether it's better to work from home or the office," *Quartz* 18 July
2017, https://qz.com/1032085/is-working-from-home-really-more-pro
ductive (accessed 2 February 2018).

31. Dan Sichel, "The productivity slowdown is even more puzzling
than you think," *Econofact* 17 May 2017, http://econofact.org/the-pro
ductivity-slowdown-is-even-more-puzzling-than-you-think (accessed 2
February 2018).

32. National Academies of Sciences, Engineering, and Medicine,
*Information Technology and the U.S. Workforce: Where Are We and
Where Do We Go from Here?* Washington, DC: The National Academies
Press, 2017, https://www.nap.edu/catalog/24649/information-tech
nology-and-the-us-workforce-where-are-we-and, https://dx.doi.org/10
.17226/24649 (accessed 2 February 2018).

33. Monideepa Tarafdar, "The dark side of digital work: How technology is making us less productive," *New Statesman* 3 February 2015,

http://www.newstatesman.com/sci-tech/2015/02/dark-side-digital
-work-how-technology-making-us-less-productive (accessed 2 February
2018).

34. Ibid.

35. C. Northcote Parkinson, "Parkinson's Law," *The Economist* 19
November 1955, http://www.economist.com/node/14116121 (accessed 2
February 2018).

36. Michael Chui, James Manyika, Jacques Bughin, et al., *The
Social Economy: Unlocking Value and Productivity Through Social
Technologies*, McKinsey Global Institute, 2012, https://www.mckinsey
.com/industries/high-tech/our-insights/the-social-economy (accessed 2
February 2018).

CHAPTER 5

1. Press Association, "Children spend only half as much time
playing outside as their parents did," *The Guardian* 27 July 2016,
http://www.theguardian.com/environment/2016/jul/27/children
-spend-only-half-the-time-playing-outside-as-their-parents-did (accessed 2 February 2018).

2. Jenny Anderson, "A study of kids' screen time explains the
vicious cycle that makes parents unable to say no," *Quartz* 1 August
2017, qz.com/1042581/a-study-of-kids-screen-time-explains-the-vicious
-cycle-that-makes-parents-unable-to-say-no (accessed 2 February 2018).

3. Amanda Lenhart, *Teen, Social Media and Technology Overview
2015*, Washington, DC: Pew Research Center, 2015. Laurel J. Felt and
Michael B. Robb, *Technology Addiction: Concern, Controversy, and
Finding Balance*, San Francisco: Common Sense Media, 2016.

4. Common Sense Media, "Landmark report: U.S. teens use
an average of nine hours of media per day, tweens use six hours,"
Common Sense Media news release 3 November 2015, http://www
.commonsensemedia.org/about-us/news/press-releases/landmark-re
port-us-teens-use-an-average-of-nine-hours-of-media-per-day (accessed
2 February 2018). "The Common Sense Census," Common Sense
Media 3 November 2015, https://www.commonsensemedia.org/the
-common-sense-census-media-use-by-tweens-and-teens-infographic
(accessed 2 February 2018).

5. Douglas A. Gentile, Olivia N. Berch, Hyekyung Choo, et al.,
"Bedroom media: One Risk Factor for Development," *Developmental*

Psychology 2017;53(12):2340–2355, https://doi.org/10.1037/dev0000399 (accessed 2 February 2018).

6. Benjamin Herold, "Can technology get kids to play outside?" *Education Week* 26 May 2015, http://blogs.edweek.org/edweek/ DigitalEducation/2015/05/technology_kids_play_outside.html (accessed 2 February 2018).

7. Colin A. Capaldi, Raelyne L. Dopko, and John M. Zelenski, "The relationship between nature connectedness and happiness: A meta-analysis," *Frontiers in Psychology* 2014;5:976, https://dx.doi.org/ 10.3389%2Ffpsyg.2014.00976 (accessed 2 February 2018). Casey Gray, Rebecca Gibbons, Richard Larouche, et al., "What is the relationship between outdoor time and physical activity, sedentary behaviour, and physical fitness in children? A systematic review," *International Journal of Environmental Research and Public Health* 2015;12(6):6455–6474, https://dx.doi.org/10.3390/ijerph120606455 (accessed 2 February 2018). Florence Williams, "This is your brain on nature," *National Geographic* January 2016, https://www.nationalgeographic.com/maga zine/2016/01/call-to-wild (accessed 2 February 2018).

8. Deborah Franklin, "How hospital gardens help patients heal," *Scientific American* 1 March 2012, https://www.scientificamerican .com/article/nature-that-nurtures (accessed 2 February 2018).

9. Bum Jin Park, Yuko Tsunetsugu, Tamami Kasetani, et al., "The physiological effects of *Shinrin-yoku* (Taking in the forest atmosphere or forest bathing): Evidence from field experiments in 24 forests across Japan," *Environmental Health and Preventive Medicine* 2010;15(1):18–26, https://doi.org/10.1007/s12199-009-0086-9 (accessed 2 February 2018).

10. Mariana G. Figueiro, Bryan Steverson, Judith Heerwagen, et al., "The impact of daytime light exposures on sleep and mood in office workers," *Sleep Health* 2017;3(3):204–215, https://doi.org/10.1016/ j.sleh.2017.03.005 (accessed 2 February 2018).

11. Uri Ladabaum, Ajitha Mannalithara, Parvathi A. Myer, et al., "Obesity, abdominal obesity, physical activity, and caloric intake in U.S. adults: 1988 to 2010," *American Journal of Medicine* 2014;127(8):717–727.e12, https://doi.org/10.1016/j.amjmed.2014.02.026 (accessed 2 February 2018).

12. Lisa Ellis, Jeffrey Saret, and Peter Weed, *BYOD: From Company-Issued to Employee-Owned Devices*: McKinsey & Co., 2012,

https://www.mckinsey.com/~/media/mckinsey/dotcom/client_service/ High Tech/PDFs/BYOD_means_so_long_to_company-issued_devices _March_2012.ashx (accessed 2 February 2018).

13. Jane McConnell, "Tracking the trends in bringing our own devices to work," *Harvard Business Review* 4 May 2016, hbr.org/2016/ 05/tracking-the-trends-in-bringing-our-own-devices-to-work (accessed 2 February 2018).

14. Intel, "Intel security study reveals millennials are more likely to unplug while on vacation than gen X," Intel Newsroom news release 21 June 2016, newsroom.intel.com/news-releases/intel-security-study -reveals-millennials-are-more-likely-to-unplug-while-on-vacation-than -gen-x (accessed 2 February 2018).

15. HomeAway, "HomeAway science of memories study," HomeAway news release 2016, http://www.homeaway.com/files/ shared/MediaCenter/vacation-equation-whitepaper.pdf (accessed 2 February 2018).

16. Stephanie Vozza, "What happens to your brain when you work on vacation," *Fast Company* 2 June 2017, http://www.fastcompany.com/ 40425251/what-happens-to-your-brain-when-you-work-on-vacation (accessed 2 February 2018).

17. Oliver Pergams and Patricia A. Zaradic, "Is love of nature in the U.S. becoming love of electronic media? 16-year downtrend in na- tional park visits explained by watching movies, playing video games, internet use, and oil prices," *Journal of Environmental Management* 2006;80(4):387–393, https://doi.org/10.1016/j.jenvman.2006.02.001 (accessed 2 February 2018).

18. Oliver R. W. Pergams and Patricia A. Zaradic, "Evidence for a fundamental and pervasive shift away from nature-based recreation," *Proceedings of the National Academy of Sciences USA* 2008;105(7):2295– 2300, https://doi.org/10.1073/pnas.0709893105 (accessed 2 February 2018).

19. *Outdoor Participation Report 2017*, Washington, DC: The Outdoor Foundation, 2017, outdoorindustry.org/wp-content/uploads/ 2017/05/2017-Outdoor-Recreation-Participation-Report_FINAL.pdf (accessed 2 February 2018).

20. Nancy Shute, "Kids are less fit today than you were back then," Shots 20 November 2013, http://www.npr.org/sections/health

-shots/2013/11/20/246316731/kids-are-less-fit-today-than-you-were-back
-then (accessed 2 February 2018).

21. Gavin Sandercock, "Poor fitness is a bigger threat to child
health than obesity," *The Conversation* 22 June 2015, theconversation
.com/poor-fitness-is-a-bigger-threat-to-child-health-than-obesity-43653
(accessed 2 February 2018).

22. G. R. H. Sandercock, A. Ogunleye, and C. Voss, "Six-year
changes in body mass index and cardiorespiratory fitness of English
schoolchildren from an affluent area," *International Journal of Obesity*
2015;39(10):1504–1507, https://doi.org/10.1038/ijo.2015.105 (accessed 2
February 2018).

23. Denis Campbell, "Children growing weaker as computers
replace outdoor activity," *The Guardian* 21 May 2011, http://www.the
guardian.com/society/2011/may/21/children-weaker-computers-replace
-activity (accessed 2 February 2018).

24. Bernd Debusmann Jr., "Many U.S. children not getting
enough exercise," Reuters 13 April 2011, http://www.reuters.com/
article/us-children-exercise/many-u-s-children-not-getting-enough
-exercise-idUSTRE73C7R920110413 (accessed 2 February 2018).

25. Tracy L. M. Kennedy, Aaron Smith, Amy Tracy Wells, et al.,
Networked Families, Washington, DC: Pew Research Center, 2008,
http://www.pewinternet.org/2008/10/19/networked-families (accessed
2 February 2018).

CHAPTER 6

1. Paul Lewis, "'Our minds can be hijacked': The tech insiders
who fear a smartphone dystopia," *The Guardian* 6 October 2017, http://
www.theguardian.com/technology/2017/oct/05/smartphone-addiction
-silicon-valley-dystopia (accessed 2 February 2018).

2. Miller McPherson, Lynn Smith-Lovin, and Matthew E.
Brashears, "Social isolation in America: Changes in core discussion networks over two decades," *American Sociological Review*
2006;71(3):353–375, https://doi.org/10.1177/000312240607100301
(accessed 2 February 2018). Keith N. Hampton, Lauren F. Sessions
Goulet, Eun Fa Her, et al., *Social Isolation and New Technology*,
Washington, DC: Pew Research Centre, 2009, http://www.pewinter
net.org/2009/11/04/social-isolation-and-new-technology (accessed

2 February 2018). Irena Stepanikova, Norman H. Nie, and Xiabin He, "Time on the Internet at home, loneliness, and life satisfaction: Evidence from panel time-diary data," *Computers in Human Behavior* 2010;26(3):329–338, https://doi.org/10.1016/j.chb.2009.11.002 (accessed 2 February 2018). Matthew E. Brashears, "Small networks and high isolation? A reexamination of American discussion networks," *Social Networks* 2011;33(4):331–341, https://doi.org/10.1016/j.socnet .2011.10.003 (accessed 2 February 2018).

    3. "Sleep and disease risk," *Healthy Sleep* 18 December 2007, http://healthysleep.med.harvard.edu/healthy/matters/consequences/ sleep-and-disease-risk (accessed 2 February 2018).

    4. Paul Lewis, "'Our minds can be hijacked': The tech insiders who fear a smartphone dystopia."

    5. Matt Richtel, "A Silicon Valley school that doesn't compute," *New York Times* 22 October 2011, http://www.nytimes.com/2011/10/23/ technology/at-waldorf-school-in-silicon-valley-technology-can-wait .html (accessed 2 February 2018).

    6. Nick Bilton, "Steve Jobs was a low-tech parent," *New York Times* 10 September 2014, http://www.nytimes.com/2014/09/11/fashion/ steve-jobs-apple-was-a-low-tech-parent.html (accessed 2 February 2018).

    7. Gregor Hasler, Daniel Buysse, Richard Klaghofer, et al., "The association between short sleep duration and obesity in young adults: A 13-year prospective study," *Sleep* 2004;27(4):661–666, https://dx .doi.org/10.1093/sleep/27.4.661 (accessed 2 February 2018). Daniel J. Gottlieb, Naresh M. Punjabi, Ann B. Newman, et al., "Association of sleep time with diabetes mellitus and impaired glucose tolerance," *Archives of Internal Medicine* 2005;165(8):863–867, https://dx.doi.org/ 10.1001/archinte.165.8.863 (accessed 2 February 2018).

    8. Julianne Holt-Lunstad, Timothy B. Smith, and J. Bradley Layton, "Social relationships and mortality risk: A meta-analytic review," *PLoS Medicine* 2010;7(7):e1000316, https://doi.org/10.1371/ journal.pmed.1000316 (accessed 2 February 2018). Pamela Qualter, Janne Vanhalst, Rebecca Harris, et al., "Loneliness Across the Life Span," *Perspectives on Psychological Science* 2015;10(2):250–264, http:// journals.sagepub.com/doi/abs/10.1177/1745691615568999 (accessed 2 February 2018).

    9. *2011 Sleep in America Poll*, Washington, DC: National Sleep

Foundation, 2011, sleepfoundation.org/sites/default/files/sleepinamer
icapoll/SIAP_2011_Summary_of_Findings.pdf (accessed 2 February
2018).

10. Katherine Sellgren, "Teenagers 'checking mobile phones in
night,'" BBC News 6 October 2016, http://www.bbc.com/news/educa
tion-37562259 (accessed 2 February 2018).

11. Accel and Qualtrics, "The millennial study," Qualtrics October
2016, http://www.qualtrics.com/millennials (accessed 2 February
2018).

12. Redhwan A. Al-Naggar and Shirin Anil, "Artificial light at
night and cancer: Global study," *Asian Pac J Cancer Prev* 2016;17(10):
4661–4664, https://dx.doi.org/10.22034/APJCP.2016.17.10.4661
(accessed 2 February 2018). N. A. Rybnikova, A. Haim, and B. A.
Portnov, "Does artificial light-at-night exposure contribute to the
worldwide obesity pandemic?" *International Journal of Obesity (Lond)*
2016;40(5):815–823, https://doi.org/10.1038/ijo.2015.255 (accessed
2 February 2018). Laura K. Fonken, Joanna L. Workman, James C.
Walton, et al., "Light at night increases body mass by shifting the
time of food intake," *Proceedings of the National Acacemy of Sciences
USA* 2010;107(43):18664–18669, https://dx.doi.org/10.1073/pnas
.1008734107 (accessed 2 February 2018). YongMin Cho, Seung-Hun
Ryu, Byeo Ri Lee, et al., "Effects of artificial light at night on human
health: A literature review of observational and experimental studies applied to exposure assessment," *Chronobiology International*
2015;32(9):1294–1310, https://doi.org/10.3109/07420528.2015
.1073158 (accessed 2 February 2018). Thabo Mosendane, Tshinakaho
Mosendane, and Frederick J. Raal, "Shift work and its effects
on the cardiovascular system," *Cardiovascular Journal of Africa*
2008;19(4):210–215, https://www.ncbi.nlm.nih.gov/pmc/articles/
PMC3971766 (accessed 2 February 2018). Céline Vetter, Elizabeth E.
Devore, Lani R. Wegrzyn, et al., "Association between rotating night
shift work and risk of coronary heart disease among women," *JAMA*
2016;315(16):1726–1734, https://doi.org/10.1001/jama.2016.4454
(accessed 2 February 2018). Fangyi Gu, Jiali Han, Francine Laden, et
al., "Total and cause-specific mortality of U.S. nurses working rotating
night shifts," *American Journal of Preventive Medicine* 2015;48(3):241–
252, https://doi.org/10.1016/j.amepre.2014.10.018 (accessed 2 February
2018).

13. Orfeu M. Buxton, Sean W. Cain, Shawn P. O'Connor, et al., "Adverse metabolic consequences in humans of prolonged sleep restriction combined with circadian disruption," *Science Translational Medicine* 2012;4(129):129ra43, https://dx.doi.org/10.1126/scitranslmed .3003200 (accessed 2 February 2018).

14. Liese Exelmans and Jan Van den Buick, "Binge viewing, sleep, and the role of pre-sleep arousal," *Journal of Clinical Sleep Medicine* 13(8):1001–1008, https://dx.doi.org/10.5664/jcsm.6704 (accessed 2 February 2018).

15. Anne-Marie Chang, Daniel Aesbach, Jeanne F. Duffy, et al., "Evening use of light-emitting ereaders negatively affects sleep, circadian timing, and next-morning alertness," *Proceedings of the National Acacemy of Sciences USA* 2015;112(4):1232–1237, https://dx.doi.org/10 .1073/pnas.1418490112 (accessed 2 February 2018).

16. Yong Liu, Anne G. Wheaton, Daniel P. Chapman, et al., "Prevalence of healthy sleep duration among adults—United States, 2014," *MMWR* 2016;65(6):137–141, http://www.cdc.gov/mmwr/vol umes/65/wr/mm6506a1.htm (accessed 2 February 2018).

17. Matthew A. Christensen, Laura Bettencourt, Leanne Kaye, et al., "Direct measurements of smartphone screen-time: Relationships with demographics and sleep," *PLoS ONE* 2016;11(11):e0165331, https:// dx.doi.org/10.1371/journal.pone.0165331 (accessed 2 February 2018).

18. Robert D. Putnam, *Bowling Alone: The Collapse and Revival of American Community,* New York: Simon & Schuster, 2001.

19. John Cacioppo and William Patrick, *Loneliness: Human Nature and the Need for Social Connection,* New York: W. W. Norton, 2008.

20. Laura Entis, "Chronic loneliness is a modern-day epidemic," *Fortune* 22 June 2016, http://fortune.com/2016/06/22/loneliness-is-a -modern-day-epidemic (accessed 2 February 2018).

21. Nick Tarver, "'Half of adults' in England experience loneliness," BBC News 18 October 2013, http://www.bbc.com/news/uk -england-24522691 (accessed 2 February 2018).

22. C. Wilson and B. Moulton, *Loneliness Among Older Adults: A National Survey of Adults 45+,* Washington, DC: AARP, 2010, assets .aarp.org/rgcenter/general/loneliness_2010.pdf (accessed 2 February 2018).

23. Julianne Holt-Lunstad, Timothy B. Smith, Mark Baker, et al., "Loneliness and social isolation as risk factors for

mortality: A meta-analytic review," *Perspectives on Psychological Science* 2015;10(2):227–237, https://dx.doi.org/10.1177/1745691614568352 (accessed 2 February 2018).

24. "Former surgeon general sounds the alarm on the loneliness epidemic," CBS News 19 October 2017, http://www.cbsnews.com/news/loneliness-epidemic-former-surgeon-general-dr-vivek-murthy (accessed 2 February 2018).

25. Sherry Turkle, *Reclaiming Conversation: The Power of Talk in a Digital Age*, New York: Penguin 2015. Sherry Turkle, *Alone Together: Why We Expect More from Technology and Less from Each Other*, New York: Basic Books, 2001.

26. David A. Baker and Guillermo Perez Algorta, "The relationship between online social networking and depression: A systematic review of quantitative studies," *Cyberpsychology, Behavior, and Social Networking* 2016;19(11): 638–648, https://dx.doi.org/10.1089/cyber.2016.0206 (accessed 2 February 2018).

27. Sebastián Valenzuela, Namsu Park, and Kerk F. Kee, "Is there social capital in a social network site? Facebook use and college students' life satisfaction, trust, and participation," *Journal of Computer-Mediated Communication* 2009;14(4):875–901, doi/10.1111/j.1083-6101.2009.01474.x/full (accessed 2 February 2018).

28. Fenne Große Deters and Matthias R. Mehl, "Does posting Facebook status updates increase or decrease loneliness? An online social networking experiment," *Social Psychological and Personality Science* 2013;4(5):579–586, https://dx.doi.org/10.1177/194855061246 9233 (accessed 2 February 2018).

29. Ethan Kross, Philippe Verduyn, Emre Demiralp, et al., "Facebook use predicts declines in subjective well-being in young adults," *PLoS ONE* 2013 Aug 14;8(8):e69841, https://doi.org/10.1371/journal.pone.0069841 (accessed 2 February 2018).

30. Brian A. Primack, Ariel Shensa, Jaime E. Sidani, et al., "Social media use and perceived social isolation among young adults in the U.S.," *American Journal of Preventive Medicine* 2017;53(1):1–8, http://dx.doi.org/10.1016/j.amepre.2017.01.010 (accessed 2 February 2018).

31. Liu yi Lin et al., "Association between social media use and depression among U.S. young adults," *Depression and Anxiety* 2016;33:323–331, http://dx.doi.org/ 10.1002/da.22466 (accessed 2 February 2018).

32. Richard Schwartz, "Triple-threat antidepressant: A morning walk with a friend," *Sunsprite* 6 October 2015, https://www.sunsprite.com/blog/triple-threat-antidepressant (accessed 2 February 2018). Tara Bahrampour, "This simple solution to smartphone addiction is now used in over 600 U.S. schools," *The Washington Post* 5 February 2018, https://www.washingtonpost.com/news/inspired-life/wp/2018/02/05/this-millennial-discovered-a-surprisingly-simple-solution-to-smart phone-addiction-schools-love-it/?utm_term=.8f5f6abf9c68 (accessed 2 February 2018).

33. Andrew K. Przybyiski and Netta Weinstein, "Can you connect with me now? How the presence of mobile communication technology influences face-to-face conversation quality," *Journal of Social and Personal Relationships* 2013;30(3):237–246, http://journals.sagepub.com/doi/abs/10.1177/0265407512453827 (accessed 2 February 2018).

34. M. A. Lapierre and Meleah N. Lewis, "Should it stay or should it go now? Smartphones and relational health," *Psychology of Popular Media Culture* 2016 April 21, https://dx.doi.org/10.1037/ppm0000119 (accessed 2 February 2018).

35. James A. Roberts and Meredith E. David, "My life has become a major distraction from my cell phone: Partner phubbing and relationship satisfaction among romantic partners," *Computers in Human Behavior* 2016;54:134–141, https://doi.org/10.1016/j.chb.2015.07.058 (accessed 2 February 2018).

36. Jean M. Twenge, "Have smartphones destroyed a generation?" *The Atlantic* September 2017, http://www.theatlantic.com/magazine/archive/2017/09/has-the-smartphone-destroyed-a-generation/534198 (accessed 2 February 2018) (accessed 2 February 2018).

37. Alexandra Samuel, "Yes, smartphones are destroying a generation, but not of kids," *JSTOR Daily* 2017 August 8, daily.jstor.org/yes-smartphones-are-destroying-a-generation-but-not-of-kids (accessed 2 February 2018).

38. Brigid Schulte, "Making time for kids? Study says quality trumps quantity," *The Washington Post* 28 March 2015, https://www.washingtonpost.com/local/making-time-for-kids-study-says-quality-trumps-quantity/2015/03/28/10813192-d378-11e4-8fce-3941fc548f1c_story.html (accessed 2 February 2018).

39. Andrew K. Przybyiski and Netta Weinstein, "Can you connect

with me now? How the presence of mobile communication technology influences face-to-face conversation quality."

40. Alexandra Samuel, "Yes, smartphones are destroying a generation, but not of kids."

41. "Social media fact sheet," Pew Research Center 12 January 2017, http://www.pewinternet.org/fact-sheet/social-media (accessed 8 February 2018).

42. Brandon T. McDaniel and Jenny S. Radesky, "Technoference: Parent distraction with technology and associations with child behavior problems," *Child Development* 2018;89(1):100–109, https://doi .org/10.1111/cdev.12822 (accessed 2 February 2018).

43. Barry Wellman, Aaron Smith, Amy Well, et al., *Networked Families*, Washington, DC: Pew Research Center, 2008, http://www .pewinternet.org/2008/10/19/networked-families (accessed 2 February 2018).

44. Robert Kraut, Michael Patterson, Vicki Lundmark, et al., "Internet paradox: A social technology that reduces social involvement and psychological well-being?" *American Psychologist* 1998;53(9):1017–1031, https://dx.doi.org/10.1037/0003-066X.53.9.1017 (accessed 2 February 2018).

45. Rebecca Maxwell, "Spatial orientation and the brain: The effects of map reading and navigation," *GIS Lounge* 8 March 2013, http:// www.gislounge.com/spatial-orientation-and-the-brain-the-effects-of -map-reading-and-navigation (accessed 2 February 2018).

46. N. Carr, "Is Google making us stupid? What the Internet is doing to our brains," *The Atlantic* July 2008, https://www.theatlantic .com/magazine/archive/2008/07/is-google-making-us-stupid/306868 (accessed 2 February 2018).

47. Ian Rowlands, David Nicholas, Peter Williams, et al., "The Google generation: The information behaviour of the researcher of the future," *Aslib Proceedings* 2208;60(4):290–310, https://doi.org/ 10.1108/00012530810887953 (accessed 2 February 2018).

48. Gary W. Small, Teens D. Moody, Prabha Siddarth, et al., "Your brain on Google: Patterns of cerebral activation during Internet searching," *American Journal of Geriatric Psychiatry* 2009;17(2):116–126, https://doi.org/10.1097/JGP.0b013e3181953a02 (accessed 2 February 2018).

49. Yifan Wang, Lingdan Wu, Liang Luo, et al., "Short-term

Internet search using makes people rely on search engines when facing unknown issues," *PLoS ONE* 2017;12(4), https://doi.org/10.1371/journal.pone.0176325 (accessed 2 February 2018).

50. Benjamin C. Storm, Sean M. Stone, and Aaron S. Benjamin, "Using the Internet to access information inflates future use of the Internet to access other information," 2017 *Memory* 25(6):717–723, https://dx.doi.org/10.1080/09658211.2016.1210171 (accessed 2 February 2018).

51. Gary W. Small, Teens D. Moody, Prabha Siddarth, et al., "Your brain on Google: Patterns of cerebral activation during Internet searching."

52. Matthew Fisher, Mariel K. Goddu, and Frank C. Keil, "Searching for explanations: How the Internet inflates estimates of internal knowledge," *Journal of Experimental Psychology: General* 2015;143(3):674–687, https://doi.org/10.1037/xge0000070 (accessed 2 February 2018).

53. E. Pariser, *The Filter Bubble: What the Internet Is Hiding from You*, London: Penguin, 2011.

54. Ashwini Nadkarni and Stefan G. Hofmann, "Why do people use Facebook?" *Personality and Individual Differences* 2012;52(3):243–249, https://www.ncbi.nlm.nih.gov/pmc/articles/PMC3335399 (accessed 2 February 2018).

55. "Distracted driving," Centers for Disease Control and Prevention 9 June 2017, http://www.cdc.gov/motorvehiclesafety/distracted_driving/index.html (accessed 2 February 2018).

56. "Largest distracted driving behavior study," Zendrive 17 April 2017, http://blog.zendrive.com/distracted-driving (accessed 2 February 2018).

57. "Texting is 23-times riskier than using cell phone while driving," *Consumer Reports* 28 July 2009, http://www.consumerreports.org/cro/news/2009/07/texting-is-23-times-riskier-than-using-cell-phone-while-driving/index.htm (accessed 2 February 2018).

CHAPTER 7

1. Adam Minter, "China's cashless revolution," *Bloomberg Businessweek* 19 July 2017, http://www.bloomberg.com/view/articles/2017-07-19/china-s-cashless-revolution (accessed 2 February 2018).

2. Maddy Savage, "Why Sweden is close to becoming a cashless economy," BBC News 12 September 2017, http://www.bbc.com/news/business-41095004 (accessed 2 February 2018).

3. Bhaskar Chakravorti, "Early lessons from India's demonetisation experiment," *Harvard Business Review* 14 March 2017, https://hbr.org/2017/03/early-lessons-from-indias-demonetization-experimento-fighting-government-corruption (accessed 2 February 2018). Feliz Solomon, "A world without cash would be good for the poor, one expert says," *Fortune* 16 January 2017, http://fortune.com/2017/01/16/cashless-rogoff-india-scandinavia (accessed 2 February 2018).

4. Ashish Malhotra, "The world's largest biometric ID system keeps getting hacked," *Motherboard* 8 January 2018, https://motherboard.vice.com/en_us/article/43q4jp/aadhaar-hack-insecure-biometric-id-system (accessed 2 February 2018).

5. David Glance, "Estonia is putting its country in the cloud and offering virtual residency," *The Conversation* 26 March 2017, theconversation.com/estonia-is-putting-its-country-in-the-cloud-and-offering-virtual-residency-75194 (accessed 2 February 2018).

6. Mark Suster, "One small change I made that improved my daily mental state," *Both Sides of the Table* 23 October 2017, bothsidesofthetable.com/one-small-change-i-made-that-improved-my-daily-mental-state-16ed8ec4b94d (accessed 2 February 2018).

7. Charles Duhigg, *The Power of Habit: Why We Do What We Do in Life and Business*, New York: Random House, 2012, https://www.penguinrandomhouse.com/books/202855/the-power-of-habit-by-charles-duhigg/9780812981605 (accessed 2 February 2018).

8. Richard H. Thaler and Cass B. Sunstein, *Nudge: Improving Decisions about Health, Wealth, and Happiness*, New Haven, Connecticut: Yale University Press, 2008, https://yalebooks.yale.edu/book/9780300122237/nudge (accessed 2 February 2018).

9. Jerald J. Block, "Issues for DSM-V: Internet addiction" (editorial), *The American Journal of Psychiatry* 2008;165(3): 306–307, https://ajp.psychiatryonline.org/doi/abs/10.1176/appi.ajp.2007.07101556 (accessed 2 February 2018).

10. Farhad Manjoo, "A charming alternative universe of you, your friends and no news," *New York Times* 17 August 2016, https://www.nytimes.com/2016/08/18/technology/a-charming-alternative-universe-of-you-your-friends-and-no-news.html (accessed 2 February 2018).

11. Stephen J. Dubner, "How to launch a behavior-change revolution," *Freakonomics* 25 October 2017, http://freakonomics.com/podcast/launch-behavior-change-revolution (accessed 2 February 2018).

CHAPTER 8

1. Siempo, "The phone for humans," Kickstarter March 2017, http://www.kickstarter.com/projects/siempo/the-phone-for-humans (accessed 2 February 2018).
2. Adam Lashinsky, "Amazon's Jeff Bezos: The ultimate disrupter," *Fortune* 3 December 2012, http://fortune.com/2012/11/16/amazons-jeff-bezos-the-ultimate-disrupter (accessed 2 February 2018).
3. "The more they play, the more they lose," *People* 10 April 2007, http://en.people.cn/200704/10/eng20070410_364977.html (accessed 2 February 2018).

CHAPTER 9

1. Jean M. Twenge, *iGen: Why Today's Super-Connected Kids Are Growing Up Less Rebellious, More Tolerant, Less Happy—and Completely Unprepared for Adulthood*, New York: Atria Books, 2017.
2. Franklin Foer, *World Without Mind: The Existential Threat of Big Tech*, New York: Penguin Press, 2017.
3. David Ginsberg, "Hard questions: Is spending time on social media bad for us?" Facebook Newsroom 15 December 2017, https://newsroom.fb.com/news/2017/12/hard-questions-is-spending-time-on-social-media-bad-for-us (accessed 2 February 2018).
4. "Salesforce CEO Marc Benioff: There will have to be more regulation . . .," CNBC 23 January 2018, https://www.cnbc.com/video/2018/01/23/salesforce-ceo-marc-benioff-there-will-have-to-be-more-regulation-on-tech-from-the-government.html (accessed 2 February 2018).
5. Dan Sabbagh, "Facebook to expand inquiry into Russian influence of [sic] Brexit," *The Guardian* 17 January 2018, https://www.theguardian.com/technology/2018/jan/17/facebook-inquiry-russia-influence-brexit (accessed 2 February 2018).
6. Noam Cohen, "Silicon Valley is not your friend," *New York Times* 13 October 2017, https://www.nytimes.com/interactive/2017/10/13/opinion/sunday/Silicon-Valley-Is-Not-Your-Friend.html (accessed 2 February 2018).

ACKNOWLEDGMENTS

We would like to thank all those who contributed to this book. It required massive amounts of research in a fast-changing technology world and was a team effort. For editing, John Harvey was instrumental in polishing prose and structuring our arguments. For research, Sachin Maini provided critical insights into our ever-morphing research objectives.

We also want to acknowledge the incredible patience and good cheer of our incredible spouses, Tavinder and Lisa, as we finished another crazy project together. They have to worry not only about technology changes but also about the roller-coaster ride their husbands took in studying them. As you will have realized, our oscillations between technology's goods and evils make Dr. Jekyll and Mr. Hyde look positively stable.

Last, we want to think Jeevan Sivasubramaniam and Neal Maillet from Berrett-Koehler for their editorial wisdom and guidance through the publishing process. Publishers don't usually provide authors with much more than deadlines and criticism; Neal and Jeevan stand out for their engagement, their enthusiasm, and their all-around understanding of authorial fumbles.

INDEX

Block, Jerald, 143
Blocking. *See also under* Auto-play
 videos
 incoming messages, 158
 notifications, 173
 phone use while driving, 131
 social-media usage, 163
 the use of applications, 156
Bloom, Nicholas, 88
Blue light
 feature to reduce exposure to,
 162
 harmful health effects of, xi, 112
 late-night exposure to, xi, 162
Books and the Internet, 64–65
Bottomless pit of content, 26–31
Bottomless well of consumption,
 26–31
Brain
 how technology changes the, 9,
 12, 71, 111, 124–128. *See also*
 Empathy
 plasticity, 12
Bright light at night, health effects of
 exposure to, 112, 113
Bring your own device (BYOD),
 100–101
Browsers, 8

Cacioppo, John, 118
Calendars, online, 163
Calendly, 163
Carr, Nicholas, 71, 125
Cashless society, 136
Casinos. *See also* Slot machine
 and addiction, 23, 40, 42–43
 methodologies used to influence
 behavior in, 21–23, 40
Cell phones. *See* Smartphone
Chat, online, 75
Chat tools, 31–32. *See also* Slack
Chief executive officers (CEOs), vii,
 100. *See also* Executive officers;
 specific CEOs
Children, xv. *See also* Parenting
 iPhones and, 36, 109, 162, 182

limiting game use by, 109–110,
 178–179, 187
protecting the privacy of, 150
restricting their use of technology,
 10, 45–46, 109–110, 156, 187–
 188. *See also* Parental controls
China, 173–174, 178
 limiting game use by, 109–110,
 178–179, 187
 moving slow, moving fast in,
 140–141
Choice, 14–15
 paradox of, 15, 54
 reduction of, xii
 restoring, 153
Circadian disruption, 112, 114. *See
 also* Sleep deprivation
Classroom, 152
 Slack and, 154
 technology in the, 79, 151–152
Cognitive biases, 45, 127
Collaboration, 82–83
Communication. *See also specific
 topics*
 decrease in one-on-one, 86
 fragmented nature of modern, 78
 instant/real-time, ix, xiii
 with loved ones using technology,
 9, 49
 scattered, unfocused, 58
Communication demands,
 prioritizing, 91
Communication responsibilities of
 workforce, decreasing the, 81
Communication tools, 31–32, 75. *See
 also specific tools*
 rise of, 77–78
Communications modes, 75. *See also
 specific modes*
Communications network, number
 of users and value of, 74
Communications overload, 73–80
Communications technology, 67. *See
 also specific topics*
 benefits of, 7, 119
 decreasing costs of, 74

ABOUT THE AUTHORS

VIVEK WADHWA is a Distinguished Fellow and professor at Carnegie Mellon University's College of Engineering. He is a globally syndicated columnist for the *Washington Post* and author of several books, including two with Alex Salkever: *The Driver in the Driverless Car: How Our Technology Choices Will Create the Future*, a 2017 finalist for the FT & McKinsey Business Book of the Year Award, and *The Immigrant Exodus*, named by *The Economist* as a Book of the Year of 2012. Wadhwa has held appointments at Stanford Law School, Harvard Law School, and Emory University, and is a faculty member at Singularity University.

Wadhwa is based in Silicon Valley and researches exponentially advancing technologies that are soon going to change our world. These advances—in fields such as robotics, artificial intelligence, computing, synthetic biology, 3-D printing, medicine, and nanomaterials—are making it possible for small teams to do what was once possible only for governments and large corporations to do: solve the grand challenges in education, water, food, shelter, health, and security. They will also disrupt industries and create many new policy, law, and ethics issues.

In 2012 the U.S. government awarded Wadhwa distinguished recognition as an Outstanding American by Choice for his "commitment to this country and to the common civic values that unite us as Americans." He was also named one of the world's Top 100 Global Thinkers by *Foreign Policy* magazine in that year. In June 2013 he was on *Time* magazine's list of Tech 40, one of the forty of the most influential minds in tech. And in September 2015, he was second on a list of "ten men worth emulating" in the *Financial Times*.

Wadhwa teaches subjects such as technology, industry disruption, entrepreneurship, and public policy; researches the policy, law, and ethics issues of exponentially advancing technologies; helps prepare students for the real world; and advises several governments. In addition to being a columnist for *The Washington Post*, he is a contributor to *VentureBeat, HuffPost*, LinkedIn's *Influencers* blog, and the American Society of Engineering Education's *Prism* magazine. Prior to joining academia in 2005, Wadhwa founded two software companies.

ALEX SALKEVER is an author, technology consultant, and senior executive. He is the co-author with Vivek Wadhwa of two books: *The Driver in the Driverless Car: How Our Technology Choices Will Create the Future* and *The Immigrant*

Exodus. Alex is also a columnist for *Fortune.* He previously served as the Technology Editor at BusinessWeek. com and as a Guest Researcher at the Duke University Pratt School of Engineering. He was most recently a Vice President at Mozilla. He speaks regularly at industry conferences, universities and schools, and to corporations and boards of directors, and serves as an adviser to several venture-backed start-ups.

Alex's speeches have covered a wide variety of topics, including the ethics of artificial intelligence, how artificial intelligence will affect the media, the future of work, the future of finance, and the impact of smartphones on our personal well-being.

As a writer for both print and online publications, Alex has penned dozens of articles exploring exponentially advancing technologies such as robotics, genomics, renewable energy, quantum computing, artificial intelligence, and driverless cars. In addition to *Fortune,* Alex's writing work has appeared in *Wired,* TechCrunch, VentureBeat, ReadWrite, *Inc.,* and *The Christian Science Monitor.* In his work as a consultant, Alex helps companies devise strategies and products to compete in the era of exponential technologies. Alex has also served on the founding teams of two software companies. He lives in the Bay Area.

Also by Vivek Wadhwa and Alex Salkever

The Driver in the Driverless Car
How Our Technology Choices Will Create the Future

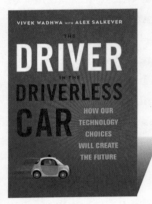

Breakthroughs such as personalized genomics, self-driving vehicles, drones, and artificial intelligence could make our lives healthier, safer, and easier. But the same technologies raise the specter of a frightening, alienating future: eugenics, a jobless economy, complete loss of privacy, and ever-worsening economic inequality. As Wadhwa and Salkever put it, our choices will determine if our future is *Star Trek* or *Mad Max*.

They offer us three questions to ask about every emerging technology: Does it have the potential to benefit everyone equally? What are its risks and rewards? And does it promote autonomy or dependence? The future is up to us to create—even if our hands are not on the wheel, we must decide the driverless car's destination.

Hardcover, 240 pages, ISBN 978-1-62656-971-3
PDF ebook ISBN 978-1-62656-972-0
ePub ebook ISBN 978-1-62656-973-7
Digital audio ISBN 978-1-62656-975-1

BK Berrett–Koehler Publishers, Inc.
www.bkconnection.com   **800.929.2929**

Berrett–Koehler
Publishers

Berrett–Koehler
Publishers

Connecting people and ideas
to create a world that works for all

Dear Reader,

Thank you for picking up this book and joining our worldwide community of Berrett-Koehler readers. We share ideas that bring positive change into people's lives, organizations, and society.

To welcome you, we'd like to offer you a free e-book. You can pick from among twelve of our bestselling books by entering the promotional code BKP92E here: http://www.bkconnection.com/welcome.

When you claim your free e-book, we'll also send you a copy of our e-newsletter, the *BK Communiqué*. Although you're free to unsubscribe, there are many benefits to sticking around. In every issue of our newsletter you'll find

- A free e-book
- Tips from famous authors
- Discounts on spotlight titles
- Hilarious insider publishing news
- A chance to win a prize for answering a riddle

Best of all, our readers tell us, "Your newsletter is the only one I actually read." So claim your gift today, and please stay in touch!

Sincerely,

Charlotte Ashlock
Steward of the BK Website

Questions? Comments? Contact me at bkcommunity@bkpub.com.

MIX
Paper from
responsible sources
FSC® C011935

Certified

Corporation
bcorporation.net